IMAGES OF WAR

BATTLESHIPS OF THE UNITED STATES NAVY

RARE PHOTOGRAPHS FROM WARTIME ARCHIVES

Michael Green

Pen & Sword

MARITIME

First published in Great Britain in 2014 by
PEN & SWORD MARITIME
An imprint of
Pen & Sword Books Ltd
47 Church Street
Barnsley
South Yorkshire
S70 2AS

ISBN 978-1-78303-035-4

Typeset by Concept, Huddersfield, West Yorkshire HD4 5JL.
Printed and bound in Malta by Gutenberg Press Ltd.

Pen & Sword Books Ltd incorporates the imprints of Pen & Sword Archaeology, Atlas, Aviation, Battleground, Discovery, Family History, History, Maritime, Military, Naval, Politics, Railways, Select, Social History, Transport, True Crime, and Claymore Press, Frontline Books, Leo Cooper, Praetorian Press, Remember When, Seaforth Publishing and Wharncliffe.

For a complete list of Pen & Sword titles please contact
PEN & SWORD BOOKS LIMITED
47 Church Street, Barnsley, South Yorkshire, S70 2AS, England
E-mail: enquiries@pen-and-sword.co.uk
Website: www.pen-and-sword.co.uk

Contents

Dedication

I would like to dedicate this book to the late Elton J. Couvillion. A US navy veteran of the Second World War, he served as a gunner's mate on board the battleship USS *Mississippi* (BB-41). He participated in the Pacific battles of Kiska Island, the Gilbert Islands, Makin Island, Marshall Islands and Palau Islands, as well as the Philippine liberation, Surigao Strait, Leyte Gulf, Lingayan Gulf and the Okinawa landing operation. He was awarded the Purple Heart for injuries suffered from enemy fire and was the sole survivor of an explosion in the No. 2 main gun turret of his ship that occurred on 20 November 1943 during the shore bombardment of enemy-occupied Makin Island.

Foreword

In 1989, while assigned to the ship's company of a nuclear-powered carrier conducting manoeuvres in the Sea of Japan, I had occasion to observe one of the last operational battleships in action. At the end of the exercise, the entire fleet gathered for the customary final formation. Although there were two immense super-carriers present, the star of the show was clearly the battleship *Missouri*. She looked as graceful and as powerful as ever and was clearly the centre of attention.

The battleship has always held a special fascination for naval enthusiasts and historians. These ships combined immense scales of armour and firepower, and occasionally speed, and represented the cutting edge of technology in their day. They were viewed with such importance that each naval power was measured by the number of battleships it possessed.

The story of American battleships is much more than the famous fast battleships that took part in the sweeping carrier raids of the Pacific War. From the first American dreadnoughts of the *South Carolina* class, American battleship designers turned out world-class ships. Innovations introduced or quickly adopted on American battleships included the concept of all-or-nothing protection (in which armour was concentrated on key parts of the ship leaving less vital areas unprotected), super-firing turrets, triple-gun turrets, oil-fired boilers and the 14-inch main gun. Generally American battleships featured excellent protection and heavy firepower but only average speed. These ships compared favourably with their British, German and Japanese counterparts.

As the major powers were all dragged into the First World War, the United States possessed the third most powerful navy in the world. During this conflict, no American battleships took part in a naval engagement with the Germans but many were active in escorting the American Expeditionary Force to Europe.

Coming out of the First World War, the United States navy was poised to grab the title as the world's most powerful by virtue of a large battleship-building programme. This was forestalled by the Washington Naval Treaty of 1922. Since the treaty prohibited the construction of new battleships, the United States navy and other navies focused on the modernization of existing ships.

The construction of American battleships resumed in 1937 with the *North Carolina* class. They were fine ships with the excellent 16-inch main gun, the best secondary gun of the war (the 5"/38), the best radar available in the world, and a fine balance of protection and speed. The following ships of the *South Dakota* class were also excellent fighting units with even more protection. The final class of battleships built in American shipyards, the *Iowa* class, was undoubtedly a graceful ship as it focused on speed over protection and firepower. These ten modern battleships led the offensive across the Pacific starting from Guadalcanal until the final raids on Japan. Supporting the newer battleships were fifteen older ships, several of which had been rebuilt following severe damage at Pearl Harbor. Instead of deciding the course of the Pacific War in a climactic duel of dreadnoughts, American battleships performed important duties as carrier escorts and as amphibious assault support ships. On two occasions American battleships took on their Japanese counterparts and both times emerged victorious. However, by war's end, the supremacy of the aircraft carrier was undeniable and soon after the war, only some units of the *Iowa* class remained active. Despite brief periods of reactivation in the Korean and Vietnam conflicts and later during the last stages of the Cold War, the day of the battleship was over.

This new book by Michael Green does a fine job in detailing this rich history. Instead of keying only on the glamorous fast battleships, this survey history covers every American battleship. Such an account does justice not only to this fading part of American naval history but also to the fine ships themselves and the men who sailed them.

Mark E. Stille
Commander, United States Navy (retired)

Acknowledgements and Notes

As with any published work, authors must depend not only on their friends for assistance but on new acquaintances made through the research for their work. Friends who provided valuable support for this book include Michael Panchyshyn, Vladimir Yakubov, Ken Estes, Randy Talbot, Martin Morgan and Mark Stille. Paul and Lorén Hannah were kind enough to provide access to their wide assortment of images of various battleships taken through their travels across the United States.

People and organizations that assisted the author in acquiring images for this book include Janice Sniker, the collection manager for the Battleship Texas, State Historic Site (SHS), as well as the helpful staffs of the Naval Historical Center, the National Archives and the Library of Congress. Other organizations from which images for this book were acquired include the Defense Visual Information Center (DVIC) and the United States Department of Defense (DOD).

Notes to the Reader

1. The US navy defines guns by battery (or use) and calibre (or size in inches or millimetres). In 'dreadnought' and 'super dreadnought' battleships these were broken down into 'main', 'secondary' and, beginning in the 1930s, 'anti-aircraft'.
2. Pre-dreadnought battleships, including second-class battleships, had a main gun battery consisting of slow-firing, large-calibre guns ranging from 10-inch up to 13-inch in bore. These were often supported by a number of faster-firing, medium-calibre guns ranging from 6-inch to 8-inch in bore. For consistency the author has decided to describe these weapons as 'intermediate' guns. Weapons smaller than 6-inch will be classified as 'secondary guns' until the 6-inch and 7-inch intermediate guns are replaced on US navy battleships in 1906 by 8-inch guns.
3. In nautical terms 'displacement' is the weight of the water displaced by a ship; this weight being equal to the weight of the vessel. According to the Washington Naval Treaty of 1922 the term 'standard displacement' refers to a ship with a full crew, minus its fuel and reserve boiler water. The tons listed under the treaty being long tons (2,400lb per ton), the US navy also refers to the displacement of its ships in 'full load tons', which is a ship with a full crew, plus all its fuel and reserve boiler water. The weight listed under full load displacement is also in long tons.

4. In nautical terms the 'main deck' of a ship is the highest complete deck (floor) extending from stem to stern and from side to side of the hull. The main deck is also typically considered a ship's 'strength deck'. A strength deck is a complete deck designed to carry not only deck loads but also hull stresses. A 'weather deck' is the uppermost deck of a ship having no overhead protection from the elements. If the weather deck is armoured it can also be referred to as a 'protective deck'. Above a ship's uppermost hull deck, the floors of a ship's superstructure are referred to as 'levels'.

Chapter One

Early Battleships

The first all metal-clad and mechanically-driven ship of the US navy debuted during the American Civil War (1861–65) and was named the *Monitor*. The ship was launched on 30 January 1862. The *Monitor* was built in response to the building of the Confederate iron-clad ship named the *Virginia*. The two ships met in battle on 9 March 1862 with inconclusive results. Despite limited success in its first battle, the US navy launched many more metal ships with different names but all were classified as 'monitors'.

Following the civil war, the US navy entered a period of serious decline as reactionaries within its officer corps refused to embrace emerging technologies such as mechanically-driven ships. Part of the problem was the high cost of coal that made wooden sailing ships more economical to operate over long distances. This self-inflicted decline in capabilities was not addressed until civilian Secretary of the Navy William H. Hunt (1881–82) and his successor William E. Chandler (1882–85) persuaded the American Congress to authorize the first steel ships of what would become known as the 'New Navy'.

In contrast to the 'Old Navy' that consisted of ocean-going wooden sailing ships and a small number of coastal waterway monitors armoured with iron plate, the New Navy consisted of modern all-steel ships. First out of the shipyards between 1884 and 1890 were the protected cruisers *Atlantic*, *Baltimore* and *Chicago*, along with the smaller gunboat, the *Dolphin*. The three cruisers were intended to operate independently as commerce raiders. In 1876 the French navy launched the first ship that was primarily made of steel.

The First US Navy Battleships

It fell to William C. Whitney, who served as Secretary of the Navy from 1885 to 1889, to begin the story of America's battleships. He convinced the American Congress to authorize (fund) the construction of one heavy armoured cruiser in 1886, later reclassified as 'second-class battleships', named the *Maine*. The second ship was named *Texas* and was classified as a battleship from the beginning. The *Maine* was an American-designed and built ship, whereas the *Texas* was based on a British ship design but constructed in an American shipyard.

Both the *Maine* and the *Texas* were commissioned by the US navy in 1895 and served as prototypes for the battleships that would follow. The *Maine* had a standard displacement of 6,682 tons and the *Texas* 6,315 tons. The *Maine* was 319 feet long and had a beam (width) of 57 feet, while the *Texas* was 308 feet 10 inches long with a beam of 64 feet 1 inch.

In-between the authorization of a ship and its commissioning are some important benchmarks. These include the date its keel (backbone) is 'laid down'. When a ship's hull and basic superstructure is completed it will then be 'launched'. Following its launching the ship is then taken to a pier for its 'fitting out', which involves adding all the components (from the propulsion system to weapons) necessary to make it functional. In the case of battleships, the fitting-out stage can take years. Once the ship is functional it is then tested by the US navy to confirm that the builder has met all the contract requirements. Once the US navy is satisfied that the builder has met all their contractual obligations, the ship is formally 'commissioned' into regular US navy service.

The main gun battery of the *Maine* was four 10-inch guns divided between two turrets, each mounting two guns. These circular main gun battery turrets were located on either side of the ship's superstructure (but not directly opposite each other) and were often referred to as 'wing turrets'. They could be fired over the front, sides and rear of the ship. In theory, they could also be fired across the deck (athwartships) through breaks in the superstructure. However, in actual practice, the firing of the main gun battery turrets athwartships led to damaging the ship's superstructure.

The intermediate gun battery of the *Maine* consisted of six pedestal-mounted 6-inch guns, four in casemates and two in open mounts. A casemate is either an armoured or unarmoured enclosure for a weapon, typically with limited traverse and elevation. The ship's secondary gun battery was made up of a number of pedestal-mounted 6-pounder (57mm) guns and two types of 1-pounder (37mm) guns arranged around the perimeter of the ship's upper decks (floors). These were generally referred to as quick-firing (QF) guns as they employed fixed rounds (with the brass cartridge case and the projectile joined together), in contrast to the separate-loading rounds of the main and intermediate battery guns on the ship.

The vertical surfaces of the main gun battery turrets on the *Maine* were protected by armour 8 inches thick. The ship's hull belt armour was 12 inches at its thickest point, with the weather/protective deck having a maximum thickness of 3 inches. The ship's conning tower was constructed of steel armour with a maximum thickness of 10 inches. The *Maine*'s underwater protection from natural or man-made dangers consisted of a double hull.

A conning tower is an armoured structure, typically located forward of a ship's unarmoured navigation bridge, housing a secondary set of bridge controls from which the ship and its weapons can be directed in battle by its command staff.

The *Texas* main gun battery consisted of two 12-inch guns. Each of these was mounted in its own circular armoured wing turret on either side of the ship's superstructure. Like the circular wing turrets on the *Maine*, the wing turrets on the *Texas* could be fired across athwartships. The intermediate gun battery on the *Texas* consisted of six pedestal-mounted 6-inch guns, four in casemates and two in open mounts. Like the *Maine*, the *Texas* also had a secondary gun battery of smaller pedestal-mounted QF 6-pounder and 1-pounder guns arranged around the perimeter of the ship's decks. Besides the various main, intermediate and secondary battery guns, both the *Maine* and the *Texas* were equipped with surface-fired torpedo tubes.

The vertical surface of the main gun battery turrets on the *Texas* was protected by armour 12 inches thick. The ship's horizontal hull belt armour was also 12 inches at its thickest point, with the weather/protective deck having a maximum armour thickness of 3 inches. The ship's armoured conning tower was 9 inches thick. Like the *Maine*, the *Texas* had a double hull for underwater protection.

Both the *Maine* and the *Texas* had a top speed of 17 knots and had triple-expansion steam engines powered by coal-fired boilers. (A knot is a unit of speed equal to one nautical mile (6,080 feet) per hour.) The *Maine* had a crew of 374 officers and enlisted men, while the *Texas* had a crew of 392. The *Maine* was sunk in an accident in February 1898 in Havana Harbor, Cuba, leading to the Spanish-American War. The *Texas* had its final decommissioning in 1911 and was later used as a target ship for many decades by the US navy. Sometimes the US navy takes ships out of service to be stored or modernized. When this occurs, they are decommissioned and when they are returned to service, they are recommissioned.

Battleship Naming Policy

The US navy policy of naming large wooden sailing ships-of-the-line after states had begun in 1817 and was later continued for the new steel battleships. The term 'ship-of-the-line' refers to the tactic practised by navies since the seventeenth century of travelling in a linear formation in order to bring the maximum number of weapons to bear on the enemy's ships. The name 'battleship' is a contraction of the phrase 'line-of-battle ship'. Battleships were also referred to as 'capital ships'.

Most American states had at least one battleship that would eventually bear their name. Some states would have more than one battleship named after them, much to the chagrin of other states' politicians. When a new battleship was assigned the name of an older vessel, its predecessor would have typically already been decommissioned and removed from US navy service. A few were reclassified and given new roles, such as a crane ship or an ammunition storage ship.

Indiana Class

Following in the footsteps of William C. Whitney in expanding the offensive might of the New Navy was Secretary of the Navy Benjamin F. Tracy, who served from 1889 to 1893. Tracy proposed to the American Congress the building of a fleet of 200 naval ships of all types. However, Congress quickly balked at the scope and cost of what Tracy envisioned. As a consolation prize, they authorized the building of three battleships in the Navy Act of 1890: the *Indiana*, the *Massachusetts* and the *Oregon*. These ships were initially classified as 'coastal' or 'coast defence' (CD) battleships as Congress was unwilling to authorize a navy that could project its offensive capabilities overseas.

With the authorization of the *Indiana*, *Massachusetts* and *Oregon*, all on the same day – 30 June 1890 – the US navy labelled them respectively battleships No. 1 through to No. 3 for record-keeping purposes. The US navy began adding the letter prefix 'USS' to its ship designations in 1907 when President Theodore Roosevelt signed an Executive Order to that effect. The three-letter prefix code stands for 'United States Ship' and has been applied to all US navy battleships mentioned in this work. A US navy ship is not assigned the prefix 'USS' until it is formally commissioned.

It was not until 17 July 1920 that the US navy adopted an arbitrary designation system that began with the letter suffix 'BB' for battleships followed by a number, which was then back-numbered to earlier ships. Therefore in 1920 the USS *Indiana* was re-designated and became BB-1, the USS *Massachusetts* BB-2 and the USS *Oregon* BB-3. The letter suffix 'BB' is not an acronym. Cruisers were assigned the letter suffix 'CC', while destroyers were assigned the suffix letters 'DD'.

The USS *Indiana* (BB-1) was commissioned on 20 November 1895; the USS *Massachusetts* (BB-2) on 10 June 1896; and the USS *Oregon* (BB-3) on 15 July 1896. The three ships were grouped together in what became known as the *Indiana* class of battleships. The US navy names its ship classes after the first ship authorized in that class by Congress.

Lessons learned from the building and service use of each class of US navy battleships were incorporated into a follow-on class in an evolutionary process with each class hopefully an improvement over the last, although this was not always the case. With a few exceptions, follow-on class battleships were generally larger and heavier than their predecessors.

The main gun battery on the three *Indiana*-class battleships consisted of four 13-inch guns divided between two circular turrets having two guns each. One of the two 13-inch gun turrets was located in front of the ship's superstructure and the other behind, referred to as 'centreline'. In nautical terms, the centreline is an imaginary straight line running the length of a ship between the bow (front) and the stern (rear). The concept of having a battleship's main guns mounted in wing turrets, as with the second-class battleship *Maine* and the *Texas*, was already obsolete by the time those ships entered US navy service.

The intermediate battery on the three *Indiana*-class battleships consisted of eight 8-inch guns divided between four circular wing turrets of two guns each, with two wing turrets on each side of the ship's superstructure. The smaller secondary battery guns ranged in size from 3-inch to 6-inch and were located in casemates on either side of the battleships. The three ships were also fitted with six surface-fired torpedo tubes. The large number of intermediate guns on the *Indiana*-class battleships reflected their higher rate of fire than the main gun batteries on ships of that time.

The vertical surfaces on the main gun battery turrets on the *Indiana*-class battleships were protected by armour 15 inches thick. The ships' hull belt armour was 18 inches at its thickest point, with the weather/protective deck having a maximum armour thickness of 3 inches. The ships' conning towers were constructed of armour 10 inches thick. Like the *Maine* and the *Texas*, the *Indiana*-class battleships had a double bottom for protection, a feature typically not seen on ships smaller than a cruiser.

The three *Indiana*-class battleships had a standard displacement ranging from 10,288 tons up to 11,688 tons. They had a length of between 350 feet 11 inches and 351 feet 2 inches, and a beam of 69 feet 3 inches. Crew complement was 473 officers and enlisted men. They had triple-expansion steam engines powered by coal-fired boilers, giving them a top speed between 15 and 16 knots. The last of the ships had their final decommissioning in 1919, following the conclusion of the First World War.

Battleship Armour

With the introduction of the battleship into US navy service, there arose an ever-increasing demand for large amounts of high-quality steel that could be forged into guns and cast or rolled into thick, resilient armour plates. As such specialized products had no other application in the United States market place at the time, American steel manufacturers had to be prodded by the government to forge steel to meet US navy specifications in order to avoid dependence upon foreign steel manufacturers. However, a series of scandals and investigations into producing what was then perceived as overly-expensive and substandard armour plate in the 1890s haunted the American steel industry. An American government attempt at building its own steel plant and undercutting the prices charged by private industry failed. Eventually the US government shared its contracts for high-quality steel with a number of commercial American firms, ensuring a broad and profitable industrial base and a steady supply of material for building additional battleships.

The *Maine* and *Texas* both employed nickel steel armour in their construction. The *Indiana*-class battleships employed both nickel steel armour and a variation of it that was face-hardened and referred to as 'Harvey armour' after its American inventor, Hayward Augustus Harvey.

Iowa Class

On the heels of the *Indiana* class of battleships, there appeared the *Iowa* class that comprised only a single ship, the USS *Iowa* (BB-4). It was authorized on 19 July 1892 and commissioned on 16 June 1897. Like the *Indiana*-class battleships, the USS *Iowa* was considered a coastal defence vessel.

The main battery of the USS *Iowa* consisted of four 12-inch guns. Like the three *Indiana*-class battleships that preceded *Iowa*, its main battery guns were mounted in circular centreline turrets located forward and aft of the ship's superstructure. The US navy labelled the centreline main battery turrets on its battleships numerically, with the foremost being turret one and then in ascending order – turret two, turret three and so forth – regardless of their location on the ship with respect to the superstructure.

The USS *Iowa*'s eight intermediate 8-inch guns were mounted in four circular wing turrets, two in each turret, with two wing turrets on either side of the ship's super-structure. There were also six casemated 4-inch guns, three on either side of the ship's hull. In addition, the ship was armed with twenty-four secondary battery guns, both in open mounts and others casemated in the ship's superstructure, plus four surface-fired torpedo tubes.

The USS *Iowa* was protected by Harvey armour, with the vertical surfaces on the main gun battery turrets protected by armour 15 inches thick. The ship's horizontal hull belt armour was 14 inches at its thickest point, with the weather/protective deck having a maximum thickness of 3 inches. The thinness of the ship's protective deck reflected the short horizontal ranges at which battleships of the day would engage each other, which meant that the problem of dealing with long-range high-trajectory plunging fire was almost non-existent. The *Iowa*'s conning tower was built from armour 10 inches thick. Like the battleships that came before it, the ship had a double hull for protection from underwater threats.

The USS *Iowa* had a standard load displacement of 11,410 tons, was 262 feet 5 inches in length and had a beam of 72 feet 3 inches. The crew complement was 486 officers and enlisted men. The ship had triple-expansion steam engines powered by coal-fired boilers, which gave it a top speed of 16 knots. Her final decommissioning took place in 1919.

Introduction to Combat

The introduction of the US navy's battleships to the demands of a wartime environment began with the sinking of the second-class battleship *Maine* on 15 February 1898 in the harbour of Havana, Cuba, which was then a Spanish colony. The ship had sailed to Cuba due to fears that a local insurrection might pose a threat to the American consulate in that city. The more scandalous elements of the American press used this sinking, initially blamed on the Spanish military, to drum up enthusiasm among the public for going to war against Spain. This was a cover for some American politicians who, driven by imperialist desires, sought an excuse to seize portions of the moribund Spanish colonial empire that interested them. The scheme worked and war was declared against Spain on 21 April 1898. Today's naval historians generally attribute the sinking of the *Maine* to a spontaneous coal bunker fire that caused a magazine (ammunition) explosion.

Not wasting any time, the US navy set about attacking and destroying the Spanish navy's mostly decrepit fleet of ships, wherever they might be found during what has become known as the Spanish-American War. The short conflict lasted from 25 April until 12 August 1898 with two major naval engagements defining this brief war. The first occurred on 1 May 1898 when a US navy squadron, minus any battleships, caught the Spanish navy Pacific fleet in Manila Harbour, the Philippines, and decimated it in a one-sided battle.

The other major sea battle of the Spanish-American War took place in the Atlantic when two US navy squadrons, which included the second-class battleship *Texas*, USS *Indiana* (BB-1), USS *Oregon* (BB-3) and USS *Iowa* (BB-4), engaged the Spanish squadron attempting to escape the harbour of Santiago, Cuba, on 3 July 1898. As in the Philippines, the Spanish ships were quickly dealt with in a lopsided battle with the loss of only one American sailor killed and approximately 100 wounded. The Spanish navy squadron lost over 2,000 men in the same engagement.

It was the faster-firing intermediate battery guns of the US navy battleships that proved to be slightly more accurate and effective than the main battery guns on the American battleships during the Spanish-American War. This led the US navy to retain the intermediate 8-inch battery guns on most of its battleships until 1908.

Kearsarge Class

Prior to the Spanish-American War, Congress authorized funding for the construction of two additional battleships in 1895. These included the USS *Kearsarge* (BB-5), the sole exception to the battleship state-naming policy. This battleship was named after a famous American steam-powered civil war 'sloop of war'. The other battleship of the *Kearsarge* class was the USS *Kentucky* (BB-6).

Technically the USS *Kearsarge* was never designated 'BB', having been decommissioned in May 1920 before the US navy began its new designation system which employed the suffix 'BB' for battleships from July 1920.

Work on the two battleships was rushed during the Spanish-American War in the hope that they might be commissioned in time to see service in that conflict but that was not to be. Both the USS *Kearsarge* and the USS *Kentucky* were launched on 24 March 1898 and commissioned in 1900.

The main battery on the two *Kearsarge*-class battleships consisted of four 13-inch guns. They were divided between two centreline circular turrets of two guns each, as was seen on the *Indiana*- and *Iowa*-class battleships that came before them. The intermediate gun battery on the two *Kearsarge*-class battleships consisted of the standard US navy 8-inch guns. However, rather than being located in circular wing turrets on either side of the ship's superstructure as on the *Indiana*-class battleships and the single *Iowa*-class battleship, they were mounted directly on top of each of the two 13-inch circular gun turrets. The secondary gun battery consisted of fourteen pedestal-mounted 5-inch guns mounted in casemates and an assortment of smaller-calibre guns. There were also four surface-launched torpedo tubes on the ships.

The two *Kearsarge*-class battleships were protected by Harvey armour, with the vertical surfaces on the main gun battery turrets protected by armour 17 inches thick. The ships' horizontal hull belt armour was 16.5 inches at its thickest point, with the protective deck armour having a maximum thickness of 3 inches. The ships' conning towers were built of armour 10 inches thick and like the preceding battleships, these ships also had double bottoms.

The two *Kearsarge*-class battleships had a full load displacement of 12,320 tons, with a length of 374 feet 4 inches and a beam of 72 feet 3 inches. Their normal crew complement was 554 officers and enlisted men. However, when operating as a flag-ship the complement would rise to 586 officers and enlisted men on both ships. The *Kearsarge*-class ships had triple-expansion steam engines powered by coal-fired boilers, providing them with a top speed of 16 knots.

The final decommissioning of the USS *Kearsarge* occurred in 1941. It was converted into a crane ship and remained in service with the US navy until 1955. The USS *Kentucky* faced its final decommissioning in 1920. Both ships were classified as coastal defence battleships.

Illinois **Class**

Three battleships of the *Illinois* class, USS *Illinois* (BB-7), *Alabama* (BB-8) and *Wisconsin* (BB-9) were authorized in 1896. The three ships were commissioned between 1900 and 1901. Like the battleships before them, the US navy classified these as coastal defence battleships.

The main gun battery on all three *Illinois*-class battleships consisted of two 13-inch gun turrets in the standard centreline configuration. Missing from these ships were the 8-inch intermediate gun batteries. Instead, each ship's remaining firepower was concentrated in its secondary gun battery that consisted of fourteen pedestal-mounted 6-inch guns located in casemates and additional small-calibre guns. There were also four surface-launched torpedo tubes on all three ships.

The three *Illinois*-class battleships were protected by Harvey armour, with the sloping vertical faces on the front of the main gun battery turrets protected by armour 16.5 inches thick. They were the first US navy battleships to dispense with the circular armoured turrets seen on all the preceding battleship classes. The horizontal hull belt armour on the *Illinois*-class battleships was 16.5 inches at its thickest point, with the protective decks having a maximum thickness of 3 inches. The ships' conning towers were constructed of armour 10 inches thick and like the battleships that came before, they had double bottoms.

The three *Illinois*-class battleships had a full load displacement of 12,150 tons and their length ranged between 373 feet 10 inches and 375 feet 4 inches. The ships had a beam of 72 feet 3 inches. Complement on the *Illinois*-class ships ranged from 531 to 536 officers and enlisted men. All three ships had triple-expansion steam engines powered by coal-fired boilers, providing a top speed of 16 knots. The final decommissioning of the three ships took place in 1920.

The design of the three battleships of the *Illinois* class was influenced by the *Majestic* class of British battleships first commissioned in 1895. At the time, the *Majestic*-class battleships were not only the largest ships of their kind but the most thickly armoured and best armed. Eventually nine ships of the *Majestic* class were constructed, making it the most numerous class of battleship ever built in any nation's navy.

Maine Class

Three battleships of the *Maine* class were authorized in 1898 in response to the Spanish-American War. Originally intended to be near copies of the *Illinois*-class battleships, it soon became apparent among the senior levels of the US navy that American battleships were far less capable than those found in some foreign navies. This resulted in some improvements to the design of the *Maine*-class battleships that included the USS *Maine* (BB-10), USS *Missouri* (BB-11) and USS *Ohio* (BB-12). All three ships were commissioned between 1902 and 1904.

The three *Maine*-class battleships had the same main gun battery of four 12-inch guns located in twin centreline gun turrets as the predecessor *Illinois*-class ships. As with the *Illinois* class, there was no intermediate 8-inch gun battery on the *Maine* class. The sixteen pedestal-mounted 6-inch secondary guns were all casemated. Instead of surface-fired torpedo tubes, the *Maine*-class battleships had submerged torpedo-launching tubes. This design feature remained so on all subsequent US navy battleship classes until the *North Carolina*-class battleships authorized in June 1936.

The armour on the sloping vertical face of the *Maine*-class battleship main gun battery turrets was 12 inches thick. The belt armour along the hull sides of the three *Maine*-class ships was 11 inches thick, with the protective deck having a maximum thickness of 3 inches. The conning towers on the ships were constructed of armour 10 inches thick. The ships had double bottoms and would be the last class of US navy battleships protected by Harvey armour.

Full load displacement of the three *Maine*-class battleships was 13,500 tons with a length of approximately 393 feet 10 inches. The beam of the ships was 72 feet 3 inches. Their crew complement was 561 officers and men. They had triple-expansion steam engines powered by coal-fired boilers, giving them a top speed of 18 knots.

Sadly, the design of the three *Maine*-class battleships was deficient and proved unpopular with the US navy because they had poor sea-keeping abilities. They had their final decommissioning between 1919 and 1922.

Virginia Class

More acceptable to the US navy were the follow-on five battleships of the *Virginia* class, of which three were authorized in 1899 and the other two in 1900. Commissioned between 1902 and 1907, these included the USS *Virginia* (BB-13), USS *Nebraska* (BB-14), USS *Georgia* (BB-15), USS *New Jersey* (BB-16) and USS *Rhode Island* (BB-17). Unlike the US navy battleships that came before them that were primarily coastal defence ships, the new *Virginia*-class battleships were authorized by a new postwar, expansionist-oriented US Congress and were designed from the beginning as ocean-going ships.

The main gun battery armament of the *Virginia*-class battleships consisted of four centreline 12-inch guns divided between two turrets with one located in front of the ship's superstructure and one behind. This was a design feature that had begun with the *Indiana*-class battleships. The intermediate 8-inch gun battery of the *Virginia*-class ships consisted of four two-gun turrets, two of which were wing turrets and the others mounted on the roofs of the two 12-inch gun turrets. This design feature was first seen on the *Kearsarge*-class battleships but not adopted on subsequent classes. The secondary battery on the *Virginia*-class battleships consisted of twelve pedestal-mounted 6-inch guns mounted in casemates.

Active duty employment of the *Virginia*-class battleships conclusively proved that mounting the intermediate 8-inch gun turrets on the larger 12-inch main gun battery turrets was unworkable and seriously degraded the effectiveness of the ship's firepower instead of aiding it. One would have thought this design fault would have been uncovered during the active service life of the preceding *Kearsarge*-class battleships.

The *Virginia*-class battleships were the first to feature an American-built version of the German-designed Krupp armour, which had already become a standard among the world's navies. Like the Harvey armour it replaced, the Krupp armour was face-hardened (also known as cemented) but in a much-improved process that greatly increased its ballistic strength. The armour on the sloping vertical face of the main gun turrets on the *Virginia*-class battleships was 12 inches thick, with the protective deck having a maximum thickness of 3 inches. The hull belt armour had a maximum thickness of 11 inches and the conning tower was constructed of armour 9 inches thick. Once again, the ships had double bottoms.

All five of the *Virginia*-class battleships had a full load displacement of 16,094 tons and a length of 441 feet 3 inches. The beam of the ships was 76 feet 3 inches. Crew complement was 812 officers and men. The *Virginia*-class battleships had triple-expansion steam engines powered by coal-fired boilers which provided them with a top speed of 19 knots. All five ships had their final decommissioning by 1920.

Showing the Flag

The new outward-looking American Congress that authorized the *Virginia*-class battleships were aided in their build-up of the US navy by President Theodore Roosevelt who came to office in 1901. Roosevelt was a strong believer in the United States becoming an important player on the world stage, using the US navy as a symbol of its military might. During his time as president, four battleships were commissioned and the keel laid for sixteen more.

Roosevelt was the man behind what is popularly referred to today by naval historians as the 'Great White Fleet' which consisted of sixteen US navy battleships circumnavigating the globe in a span of fourteen months, starting in 1907 and ending in 1909. The name for this sailing came from the white hulls and buff upper works of the US navy ships that took part. This paint scheme was the standard for US navy ships until 1914, except for the wartime grey applied during the Spanish-American War, and after 1914.

Between 1907 and 1909 the Great White Fleet sailed approximately 43,000 miles and made numerous ports of call on a variety of continents, in what was thinly-veiled gunboat diplomacy. Besides showing off the military muscle of the US navy to potential naval competitors, especially Japan, it also provided an opportunity for the service to evaluate its ability to supply its ships in the far corners of the globe. Lessons learned from the deployment were incorporated into later US navy battleship designs.

Connecticut Class

The follow-on to the battleships of the *Virginia* class were the six battleships of the *Connecticut* class that included the USS *Connecticut* (BB-18), USS *Louisiana* (BB-19), USS *Vermont* (BB-20), USS *Kansas* (BB-21), USS *Minnesota* (BB-22) and the USS *New Hampshire* (BB-25). They were authorized between 1902 and 1906. The first to be commissioned was the USS *Connecticut* in September 1906 and the last the USS *New Hampshire* in March 1908.

The number assigned to the battleship *New Hampshire* (BB-25) was out of sequence with the other ships in its class as two battleships of the *Mississippi* class, the USS *Mississippi* (BB-23) and the USS *Idaho* (BB-24) were authorized in March 1904, prior to its authorization. The *Mississippi* and *Idaho* were both commissioned in 1908.

The main gun battery on the six *Connecticut*-class battleships consisted of four centreline 12-inch guns divided between two turrets, one forward and one aft of the ship's superstructure, each with two guns. The intermediate gun battery on the *Connecticut* class included eight 8-inch guns divided between four wing turrets, two on either side of the ships' superstructures, and twelve casemated 7-inch guns, six on either side of the ships' hulls. The secondary battery guns mounted in the ships' superstructures ranged from 3-inch guns to .30 calibre machine guns.

The armour on the sloping vertical turret face of the *Connecticut*-class battleships' main gun batteries was 12 inches thick. The hull belt armour on the ships had a maximum thickness of 11 inches, with the protective deck having a maximum thickness of 3 inches. The conning towers on the ships were built of 9-inch thick armour. The ships had double bottoms.

The *Connecticut*-class battleships earned a favourable reputation in active service as being more seaworthy than the *Virginia* class and more economical to operate. The operational improvements to the battleships came about because of the pioneering maritime research work done by American naval architect and engineer David W. Taylor of the US navy. It was he who devised a large tank in which ship models had their hull hydrodynamic (sea-handling) characteristics tested before designs were finalized for their full-sized counterparts. From Taylor's ground-breaking work it was discovered that a long hull form operated more efficiently at high speed in the water than a shorter one. As a result, the *Connecticut*-class battleships were longer than the *Virginia* class. Another factor in early battleship classes being relatively short in length was the perpetual parsimony of the American Congress. Funding was kept to a minimum, so ships had to be compact due to cost.

Displacing full load of 17,650 tons, the six *Connecticut*-class battleships had a length of 456 feet 4 inches and a beam of 76 feet 10 inches. Crew complement was 827 officers and men. The *Connecticut*-class ships had triple-expansion steam engines powered by coal-fired boilers, giving them a top speed of 18 knots. All six ships had their final decommissioning by 1923.

Mississippi Class

The two *Mississippi*-class battleships had a smaller full load displacement than the *Connecticut* class, at only 14,465 tons. Their length was 382 feet with a beam of 77 feet. Being lighter and shorter than their predecessors they were cheaper to build. The more affordable cost of the *Mississippi*-class battleships was an important consideration for a then very budget-minded Congress. The downside of these two smaller, more affordable battleships was that they proved far less capable than *Connecticut*-class battleships in their sea-handling characteristics.

The armament arrangement on the two *Mississippi*-class battleships was similar to that of the *Connecticut* class with four 12-inch main battery guns, divided between two turrets of two guns each, with one forward of the ship's superstructure and one aft. There were eight 8-inch intermediate battery guns, divided between four wing turrets armed with two guns each, two on either side of the ship's superstructure. The intermediate gun battery also included eight 7-inch casemated guns, four on either side of the hull, four less than the *Connecticut*-class battleships.

The secondary battery guns on the *Mississippi*-class ships consisted of everything from 3-inch guns down to .30 calibre machine guns arrayed around the ships' super-structure. These two battleships proved to be the last in the US navy to be fitted with ram bows. The last effective use of the ram bow had occurred at the Battle of Lissa in 1866 between the Austrian and Italian fleets.

The sloping vertical turret faces of the ships' main gun batteries were protected by 12 inches of steel armour, with the hull belt armour having a maximum thickness of 11 inches. The maximum thickness of the protective deck was 3 inches, while the conning tower armour was 9 inches thick. The ships, as their predecessors, had double bottoms.

Complement of the two *Mississippi*-class battleships was 744 officers and men. They had triple-expansion steam engines powered by coal-fired boilers, providing a top speed of 17 knots. The ships were considered by the US navy to be great failures in design and resulted in them being in service for only six years before having their final decommissioning in 1914. They would go on to have a second life in the Greek navy until sunk by German aircraft in 1941 during the early stages of the Second World War.

An engraving of the first sea battle between metal-clad ships which took place during the American Civil War on 9 March 1862 near Hampton Roads, Virginia and included the Union Navy's steam-powered *Monitor* (on the left) and the Confederate Navy's steam-powered ship, the *Virginia* (on the right). The *Virginia* was built upon the hull of the captured Union Navy ship the *Merrimack* but the correct Confederate name was CSS *Virginia*. The engagement between the two ships resulted in a tactical draw as neither could sink the other with its onboard weapons. (*National Archives*)

The Union Navy's *Monitor* seen here with damage from its battle with the *Virginia* was designed by civilian engineer and inventor John Ericsson. Unlike any ship before it, the *Monitor* had a turret capable of traversing 360 degrees and was armed with two 11-inch muzzle-loading smoothbore guns, protected by 8 inches of armour. These ships were often called 'ironclads', a reference to the material employed in armouring them. (*National Archives*)

Following on the heels of the *Monitor*, during the American Civil War the Union Navy built many more all-metal steam-powered ships in a number of configurations. Pictured is the *Onondage* on the James River in Virginia in 1864. The tarps seen on the ship indicate the crew is not anticipating combat. Rather than the single 360° traversing turret of the *Monitor*, the *Onondage* had two: one armed with two 15-inch muzzle-loading smoothbore guns and the smaller with two 8-inch versions of the same weapon. (*National Archives*)

Even as the US navy committed itself to the building of battleships in 1890, it continued to authorize the construction of a small number of now steel armour-protected monitors such as the USS *Ozark* (BM-7) seen here, originally named the *Arkansas* until 1909. The coal-fired monitor was commissioned in 1902 and performed battleship-like roles. Armament was a 360° traversing turret with two 12-inch breech-loading rifled guns. The USS *Ozark* was decommissioned in 1919. (*National Archives*)

Looking much like a small battleship is the coal-fired monitor USS *Cheyenne* (BM-10), originally named the *Wyoming* and commissioned in 1902. The *Cheyenne* was converted into a submarine tender in 1913 and remained in service as a training ship with the US navy until decommissioned in 1926 and sold for scrapping in 1939. Like all monitors, it had a very low free-board and was therefore not intended for open ocean operations. (*National Archives*)

The coal-fired *Maine* was originally classified as a heavy armoured cruiser and then reclassified as a second-class battleship by the US navy. The American-designed ship was constructed by the New York Navy Yard, Brooklyn, New York. It was built in response to a naval build-up among South American countries that left the US navy at a perceived disadvantage. In December 1897 it was assigned to the US navy's North Atlantic Squadron. (*National Archives*)

Due to the fear of armed conflict in the Spanish colony of Cuba, the American government had the US navy dispatch the second-class battleship *Maine* to safeguard American interests on the island. Pictured is the ship entering Havana Harbor on 25 January 1898. The immaturity of American industry in the late 1800s meant that it took seven years to complete the *Maine* and resulted in the ship being obsolete, either as a cruiser or battleship, by the time it was commissioned. (*National Archives*)

Three weeks after its arrival in Havana, Cuba, on the evening of 15 February 1898 the *Maine* was rocked by a massive explosion that destroyed the forward third of the ship. The remainder of the vessel quickly settled on the bottom of Havana Harbor as seen here. Because the enlisted personnel's quarters were located in the forward part of the ship, 266 men perished in the blast. The explosion was blamed on Spanish military action at the time and helped lead to the Spanish-American War two months later. (*National Archives*)

Unlike the American-designed second-class battleship *Maine*, the second-class battleship *Texas* shown here was based on a British design. It was constructed by the Norfolk Navy Yard, Portsmouth, Virginia and saw action during the 3 July 1898 Battle of Santiago de Cuba when the Spanish fleet based in Cuba was blasted into oblivion by the blockading US navy squadron. *Texas* later served with the US navy Atlantic Squadron until 1908 and its name was assigned to a new battleship in 1911. (*National Archives*)

This view of the second-class battleship *Texas* shows one of the ship's two main gun battery turrets, each armed with a single 12-inch gun, which projected outward from the hull and weather deck. The main armour belts on the sides of the ship's hull were 12 inches thick. The battleship's weather/armoured deck ranged in thickness from 1 to 3 inches, with its conning tower protected by a maximum armour thickness of 12 inches. (*Library of Congress*)

On the superstructure of the *Texas*, manned by a US Marine, is one of the four 1.5-inch Hotchkiss five-barrel hand-cranked revolving guns that the ship carried as an anti-torpedo boat weapon. The ship was also armed with other anti-torpedo guns including twelve 2.24-inch guns originally referred to as 'six-pounders' from the weight of their projectiles. They were later assigned a Mark (Mk) number based on when they entered US navy service. (*Library of Congress*)

The first ship designed from the beginning as a battleship for the US navy was the coal-fired USS *Indiana* (BB-1) shown here. It was built by William Cramp & Sons, Philadelphia, Pennsylvania, minus the guns and armour. The battleship's four 13-inch main guns were divided between two turrets with two guns each, one located forward of and the other aft of the ship's superstructure in the centreline position. The main advantage of this arrangement was that all the ship's main guns could be fired in a broadside. (*National Archives*)

Forming part of the three-ship *Indiana*-class of battleships was the USS *Massachusetts* (BB-2), shown here with smoke exiting its two stacks, sometimes referred to by the US navy as 'smoke pipes' in the past. In the British Royal Navy, the stacks are referred to as funnels. Like the USS *Indiana*, the USS *Massachusetts* was constructed by William Cramp & Sons, except for its guns and armour. The large pole above the ship's superstructure mast was then referred to as a 'military mast' in the US navy. (*National Archives*)

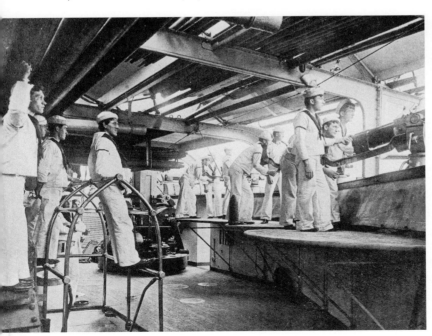

This picture taken on board the USS *Massachusetts* shows crew members manning some of the 2.24-inch guns during a training exercise. The only protection afforded to the gun crews came from the armoured parapet that they stood behind. The 2.24-inch guns formed part of the secondary battery on US navy battleships of the day and were intended to ward off torpedo boat attacks. One of the ship's small boats can be seen stowed directly above the guns. (*Library of Congress*)

The USS *Massachusetts* and her sister ships had coal-fired boilers. This photograph shows the no doubt extremely hot and uncomfortable boiler room of the battleship. The seamen in the picture are referred to as 'firemen' and it was their job to shovel the coal in the foreground into the boilers. They were also responsible for raking out the ash pits located underneath the boilers and disposing of the ashes over the sides of the ship while at sea. In the Royal Navy the firemen are referred to as stokers. (*Library of Congress*)

(*Opposite*) Sailors on board the USS *Massachusetts* are drinking or waiting for their daily ration of distilled spirits (alcohol). This practice was ended in February 1899 and reflected a growing anti-alcohol movement within the United States. Alcohol was officially banned by the American government between 1920 and 1933, an era known as Prohibition. Naval officers were allowed a wine mess on their ships until that practice was ended in June 1914. The officers were once again allowed to have alcohol on their ships beginning in 1934 but not so the enlisted personnel. (*Library of Congress*)

(*Above*) Part of the US Marine Corps contingent aboard the USS *Massachusetts* pose for the photographer in their dress uniforms. The Marines performed the role of policemen aboard ships to ensure discipline and manage the brig (prison). They were also called upon to operate some of the ship's secondary guns, or sometimes formed into landing parties to go ashore when needed. (*Library of Congress*)

The petty officers' mess (dining area) of the USS *Massachusetts* appears in this picture. The service stripes hash marks seen on the petty officers in the foreground represent years of service; one for each four years. They were red on blue uniforms and blue on white uniforms. Visible in the left rear background is one of the battleship's 18-inch torpedo tubes. (*Library of Congress*)

Bunks were unknown on early US navy ships for enlisted personnel. They made do with hammocks, as seen here on the USS *Massachusetts*. The ship had its final decommissioning in March 1919 and was later scuttled off the coast of Pensacola, Florida, after use as a target ship. The *Massachusetts* was never officially scrapped and in the 1950s it was turned over to the state of Florida. It now serves as an artificial reef and a diving spot. (*Library of Congress*)

The third battleship of the *Indiana*-class was the USS *Oregon* (BB-3), constructed by the Union Iron Works, San Francisco, California. It became famous for its sixty-six-day and 16,000-mile transit from San Francisco, California to the state of Florida to take part in the Spanish-American War, before the Panama Canal existed. Due to its then remarkable journey and its part in the Spanish-American War, the *Oregon* acquired the popular nickname 'Bulldog of the Navy'. *(National Archives)*

A young Marine on the USS *Oregon* mans a 1.5-inch gun, also referred to as a 'one-pounder'. It formed part of the ship's secondary battery and was employed as an anti-torpedo boat weapon. The first steam-powered torpedo boat was British and appeared in 1877. The US navy placed thirty-five torpedo boats into service between 1890 and 1902. The first successful sinking of a ship by a torpedo boat occurred during the Russo-Turkish war of 1877–78. Torpedo boats eventually evolved into what became known as destroyers. *(Library of Congress)*

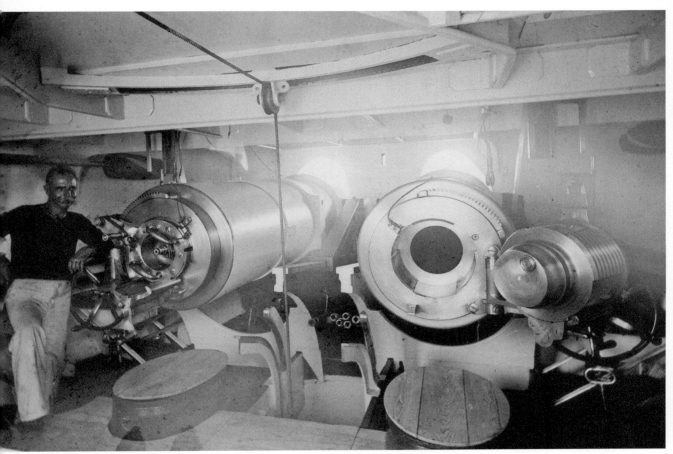

This photograph was taken inside one of the two main gun battery turrets of the USS *Oregon* with a sailor proudly standing next to one of the large twin 13-inch guns. It was originally hoped that the *Indiana*-class battleships could be fitted with 16-inch main guns but, as this was beyond the capabilities of American industry at the time, the 13-inch main guns eventually mounted on the ships were the back-up choice. (*Library of Congress*)

The officers of the USS *Oregon* and a guest in civilian clothing pose for a picture in their mess. Little thought must have been given to the risk of fire as demonstrated by the wooden panelling along the walls and the wooden table and chairs. The ship was placed into reserve in 1911 but was brought back into service in 1917 when the United States officially joined in the First World War against Germany. (*Library of Congress*)

A close-up photograph of the forecastle deck (the forward section of the weather deck) of the USS *Oregon* with the forward main gun battery turret visible, as well as the smaller intermediate battery wing turrets behind it on either side. Rising up from behind the main gun battery turret is the foremast that has a number of platforms for the mounting of searchlights for night-fighting. The open navigation bridge visible behind and above the main gun battery turret is typically referred to as a 'flying bridge'. (*National Archives*)

The USS *Oregon* in dry dock was now fitted with a very tall cage mast behind the ship's superstructure. A military mast has been retained at the front of the ship, behind the superstructure. Like the other two *Indiana*-class battleships, it was decommissioned and later recommissioned on a number of occasions, eventually being donated to the state of Oregon in 1925 as a memorial. With the outbreak of the Second World War, the ship was sold for scrap.

(National Archives)

Shown in dry dock is the coal-fired USS *Iowa* (BB-4), the only ship in its class. Very evident is the ram bow on the front of the ship's hull, an ancient maritime weapon that many in the US navy still believe had some validity in combat. The *Iowa* was built by William Cramp & Sons, Philadelphia and saw action during the Spanish-American War. (*National Archives*)

The captain of the USS *Iowa* poses for the photographer with the stern look of one who is comfortable with the reins of command. The officers who assumed the leadership roles on US navy battleships were generally a very conservative-oriented body who vested their careers in the status quo, with little inclination to embrace any new technology or innovation that might challenge their position within the service. (*Library of Congress*)

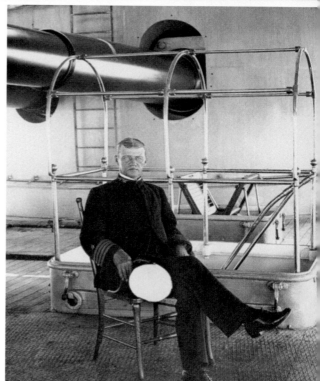

A group of sailors and enlisted Marines appear in this group photograph taken on board the USS *Iowa*. They are posing in front of the forward main gun battery turret. Visible behind and above the forward main gun turret is the wooden navigation bridge, which formed part of the superstructure bridge assembly and was the location from which an officer controlled the ship. The narrow walkway extending out from the navigation bridge, matched by one on the other side of the bridge, was referred to as a 'bridge wing' and was used by the battleship's officers when docking or manoeuvring. (*Library of Congress*)

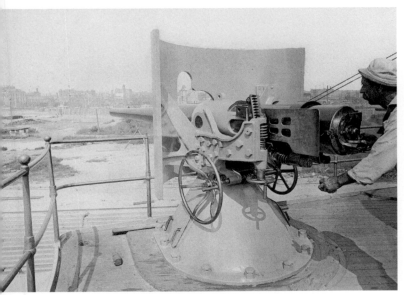

Taken on board the USS *Iowa*, this photograph shows a close-up view of an armoured shield-protected 4-inch gun that was the mainstay of early US navy battleships' secondary batteries. With the operational range of the torpedo being constantly increased, the 4-inch guns on early US navy battleships were soon outranged and later removed. Their replacement, from 1910 onwards, was a larger and longer-ranged 3-inch gun. (*Library of Congress*)

This photograph is of the rear main gun battery turret on the USS *Iowa*. Visible on the ship's rear superstructure are some of the secondary battery guns. As these secondary battery guns were primarily anti-torpedo boat weapons, they tended to be mounted as high as possible on a ship's superstructure to attain the best possible field of fire. (*Library of Congress*)

Being moved on the weather/armoured deck of the *Iowa* is a Model 1894 Driggs-Schroeder 6-pounder (57mm) rapid-fire gun. It was intended for use by the ship's crew when going ashore during wartime or to show the flag. The use of sailors as infantry and artillerymen by the US navy was very common during the nineteenth century. (*Library of Congress*)

Among the many types of weapons carried on board early US navy battleships for defence against torpedo boat attacks was the machine gun. Pictured on the USS *Iowa* is a sailor posing with a M1895 Colt-Browning machine gun, more commonly known as the 'potato digger' because of its unique, forward-mounted operating lever. The weapon was designed by American inventor John M. Browning and in US navy service fired a 6mm round. (*Library of Congress*)

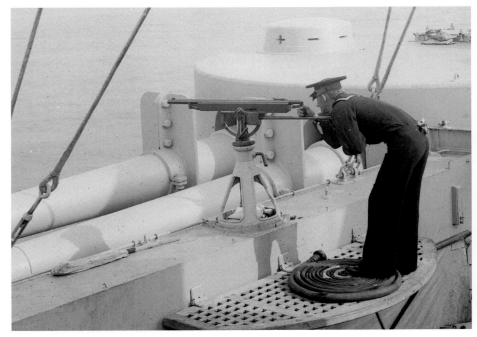

This picture of the USS *Iowa* shows that its rear military mast was replaced by a much larger, taller cage mast. At the top of this mast is a small compartment that contained various pieces of equipment used to control the ship's main gun and intermediate gun batteries. The cage masts appeared on US navy battleships from 1910 during a modernization process. The *Iowa* was sunk as a target ship by the US navy in March 1923. (*National Archives*)

Steaming at speed, the coal-fired USS *Kearsarge* (BB-5) was built by the Newport News Shipbuilding Company, Newport News, Virginia. It sailed with the Great White Fleet between 1907 and June 1909 and then served as a training ship during the First World War. After its final decommissioning in 1920, its main and intermediate guns were fitted onto later US navy battleships and cruisers while its hull was used as the basis for a mobile, heavy-lift crane ship until being sold for scrap in 1955. *(National Archives)*

The coal-fired USS *Kentucky* (BB-6), also built by the Newport News Shipbuilding Company, formed the second ship of the *Kearsarge* class and, like the *Kearsarge*, the intermediate gun battery turrets were located directly on top of the main gun battery turrets, doing away with the wing battery turrets seen on its predecessors. The secondary battery guns are arranged like those of a wooden sailing ship-of-the-line, in linear fashion along the side of the hull. *(National Archives)*

Befitting his rank, the captain of the USS *Kentucky* poses in his well-appointed quarters which include leather-covered chairs and sofa. Like the USS *Kearsarge*, the *Kentucky* sailed around the world's oceans with the Great White Fleet between 1907 and 1909. The ship was stationed off the Mexican coast between 1915 and 1916 to protect American interests during the Mexican Revolution that lasted from 1910 to 1920. (*National Archives*)

The rear main gun battery turret on the USS *Kentucky* with an intermediate battery gun turret mounted on top of it. It was an unusual design feature that not everybody believed had been well thought-out and was not repeated on the two follow-on classes of battleships. The armour protection on the intermediate gun battery turret ranged from 6 to 11 inches in thickness. The ship was sold for scrapping in 1924. (*National Archives*)

The coal-fired USS *Illinois* (BB-7) was the first of three battleships of the *Illinois* class. Again it was built by the Newport News Shipbuilding Company. Identifying features of the *Illinois* class were the lack of any intermediate gun battery turrets, and its two side-by-side stacks instead of the more typical arrangement with one stack behind the other as seen on previous US navy battleship classes. The *Illinois* had its final decommissioning in 1920 but was not sold for scrapping until 1956. (*National Archives*)

Built by William Cramp & Sons, Philadelphia, the coal-fired USS *Alabama* (BB-8) formed part of the *Illinois* class of three battleships. It sailed with the Great White Fleet between 1907 and 1909 but during the First World War it served only as a training ship. The final decommissioning took place in 1920. It was later used as a target ship in 1921 and what remained was sold for scrap in 1924. (*National Archives*)

Riding at anchor is the coal-fired USS *Wisconsin* (BB-9) constructed by the Union Iron Works, San Francisco. Like her two sister ships of the *Illinois* class, the *Wisconsin* sailed with the Great White Fleet and later became part of the US navy Reserve Atlantic Fleet. Thereafter it served as a training ship until its final decommissioning in 1920 and eventual scrapping in 1922. (*National Archives*)

In honour of the second-class battleship *Maine*, William Cramp & Sons of Philadelphia constructed the coal-fired USS *Maine* (BB-10) for the US navy. It was the first of three battleships of the *Maine* class. Like the *Illinois* class that preceded it, they lacked any intermediate battery gun turrets. An identifying feature of the *Maine*-class battleships was the arrangement of three stacks one behind the other on the centreline, as shown here. (*National Archives*)

Pictured is one of the two 18-inch torpedo-launching tubes on the USS *Maine*. The ship sailed with the Great White Fleet between 1907 and 1908, being detached early due to her heavy coal consumption. As with other US navy battleships of the era, it served as a training ship during the First World War. It had its final decommissioning in 1920 and was scrapped in 1922. (*Library of Congress*)

The second ship of the three *Maine*-class battleships was the coal-fired USS *Missouri* (BB-11), shown here sailing in formation with another battleship in line behind it. It was built by the Newport News Shipbuilding Company, Virginia. Notice that both of the ship's original military masts have been replaced by cage masts. The USS *Missouri* had its final decommissioning in 1919 and was sold for scrapping in 1922. (*National Archives*)

Part of the trio of *Maine*-class battleships was the coal-fired USS *Ohio* (BB-12) seen here. Constructed by the Union Iron Works, the ship sailed with the Great White Fleet from 1907 until 1909 and like its sister ships saw its final role as a training ship during the First World War. It was decommissioned in 1922 and sold for scrapping in 1923. (*National Archives*)

In a very non-warlike pose, the USS *Ohio* is covered with the crew's laundry drying out. The armour along the ship's hull sides was from 7 to 11 inches thick, while the protective deck armour was between 2.5 and 4 inches deep. The sixteen casemated 6-inch secondary battery guns were protected by a maximum of 6 inches of armour. (*National Archives*)

The coal-fired USS *Virginia* (BB-13) was constructed by the Newport News Shipbuilding Company and was the lead for the five ships that constituted the *Virginia* class of battleships. It is seen here in 1905 being tested by the US navy prior to being commissioned. Notice that the majority of its weapons are not yet fitted, including its hull-mounted casemated 3-inch guns and two of its intermediate 8-inch gun turrets that would later be mounted one on top of each of the two main gun battery turrets. (*National Archives*)

Built by the Moran Brothers, Seattle, Washington, was the coal-fired USS *Nebraska* (BB-14), the second ship in the *Virginia* class. As with the *Maine*-class battleships that preceded it, there were three centreline stacks. After sailing with the Great White Fleet, the ship spent time with the Atlantic Fleet and then served off the Mexican coast in both 1914 and 1916 to protect American interests during the Mexican Revolution. During the latter part of the First World War the *Nebraska* undertook convoy escort duties in the Atlantic. (*National Archives*)

Pictured is the USS *Nebraska*. The armour thickness along the sides of the *Virginia*-class battleship's hull ranged from 6 to 11 inches. The armoured conning towers, located behind the forward main gun battery turret and underneath the semi-enclosed navigation bridges, had a near-duplicate set of controls that were present on their navigation bridges. The conning towers were protected by 9 inches of armour. (*National Archives*)

Sporting a dazzle camouflage paint scheme during the First World War is the USS *Nebraska*. The purpose of this strange paint scheme, invented by the British, was to make it difficult for the coincidence range-finders on enemy submarines or ships to estimate a ship's range, as well as speed and heading. The practice was adopted by the US navy in 1918 and employed in a more subdued form on some US navy battleships during the Second World War. (*National Archives*)

The USS *Georgia* (BB-15) was part of the *Virginia* class and constructed by the Bath Iron Works, Bath, Maine. Like its counterparts from the same era, it sailed with the Great White Fleet and then spent time with the Atlantic Fleet. During the First World War she served as a training ship and eventually performed some convoy escort duties. At the end of the conflict, the ship transported American soldiers from France back to the United States. (*National Archives*)

Constructed by the Fore River Company, Quincy, Massachusetts is the coal-fired USS *New Jersey* (BB-16), part of the *Virginia* class. In this picture the ship is sporting a First World War-era dazzle camouflage paint scheme. During the conflict it served as a training ship and then aided in returning American soldiers from France back to the United States following the end of the fighting. The decommissioned ship was sunk by American aircraft during a test of their effectiveness on 5 September 1923. (*National Archives*)

The USS *Rhode Island* (BB-17) was also built by the Fore River Company, Massachusetts. Its service life mirrored that of the other four *Virginia*-class battleships, including forming part of the Great White Fleet and then sailing with the Atlantic Fleet. It also spent time off the Mexican coast during the Mexican Revolution. In reserve before the First World War, it was brought back to service during the conflict for training duties and anti-submarine patrol duty off the Maryland coast. (*National Archives*)

The follow-on to the five *Virginia*-class battleships was the six coal-fired *Connecticut*-class battleships: the USS *Connecticut* (BB-18) pictured here, the USS *Louisiana* (BB-19), USS *Vermont* (BB-20), USS *Kansas* (BB-21), USS *Minnesota* (BB-22) and USS *New Hampshire* (BB-25). Notice that the designers of this battleship class had learned their lessons and did away with the intermediate gun battery turret mounted on the top of the main gun battery turrets. The intermediate wing gun battery turrets were retained on this class of battleships. (*National Archives*)

Unlike the USS *Connecticut* built at the New York Navy Yard, Brooklyn, New York, the USS *Louisiana* shown here was constructed at the Newport News Shipbuilding Company, Virginia. The high point of the ship's service life was transporting American President Theodore Roosevelt to Panama to see the work being done on the Panama Canal, which he had championed. The ship was sold for scrapping in 1923. (*National Archives*)

The third *Connecticut*-class battleship, the USS *Vermont*, is shown here. The ship was built at the Fore River Company, Massachusetts and would see the normal interwar and then wartime duties as a training ship before being sold for scrapping in 1923. *(National Archives)*

Another *Connecticut*-class battleship was the USS *Minnesota* seen here in this stern view. It was built by the Newport News Shipbuilding Company, Virginia. Like the other battleships in its class, its hull was protected by a belt of armour ranging from 6 to 11 inches thick. Late in the First World War the ship was damaged by a mine laid by a German submarine. The *Minnesota* was scrapped in 1924. *(National Archives)*

A sister ship of the USS *Vermont* was the USS *Kansas*, pictured here. It was constructed at the New York Shipbuilding Corporation, Camden, New Jersey. Her service life was uneventful, including time spent as a training ship during the First World War. Unlike many of her counterparts, the ship spent time in the Pacific Ocean following the First World War and was then sold for scrapping in 1923. (*National Archives*)

The last ship authorized in the *Connecticut*-class of battleships was the USS *New Hampshire*, constructed by the New York Shipbuilding Corporation, New Jersey. It is shown here during a trial run prior to its main gun battery turrets being fitted. Also missing are the guns of the intermediate wing turrets. The *New Hampshire* was sold for scrapping in 1923. (*National Archives*)

In-between the authorization of ships five and six of the *Connecticut*-class battleships, the American Congress authorized the building of two smaller and lighter *Mississippi*-class battleships. These comprised the USS *Mississippi* (BB-23) shown here and the USS *Idaho* (BB-24), both of which were built by William Cramp & Sons, Philadelphia. (*National Archives*)

Pictured is the USS *Idaho* in US navy service before it and its sister ship, the USS *Mississippi* were sold to the Greek navy in 1914. In Greek navy service the *Idaho* became the *Lemnos* and the *Mississippi* was renamed the *Kilkis*. Both ships saw limited service in the First World War with the Greek navy and became training ships by the 1930s. Both battleships were sunk early in the Second World War by German aircraft. They were raised in the 1950s and then scrapped. (*Maritime Quest*)

Chapter Two

Dreadnoughts

In the early 1900s a number of developments began to influence the design of battleships around the globe. One of these was the idea among various far-sighted naval ordnance experts of doing away with the intermediate gun batteries on battleships and increasing the number of main guns. This was being considered for a number of reasons. The larger and heavier the projectile being fired, the more accurate and more destructive it was compared to its smaller counterparts which could no longer be counted on to possess the kinetic energy needed to penetrate the improved armour plate then being mounted on newer ships.

An additional reason for the growing acceptance of dispensing with intermediate gun batteries on battleships and increasing the number of main gun batteries included the fact that the rate of fire of the larger guns was becoming closer to that of the intermediate gun batteries. The ever-increasing range of torpedoes also compromised the stand-off range once provided by the battleship's intermediate and secondary gun batteries, making them somewhat superfluous. An added benefit of doing away with the intermediate batteries' guns was the logistical simplification of having to supply battleships with one less type of ammunition. Another big reason for removing the intermediate batteries was that it made it easier to track the fall of shot for the main gun batteries.

Another change to battleship design in the early 1900s was the improvement in fire control that allowed the main gun battery to be employed at longer ranges. Prior to this time the lack of an effective fire-control system on battleships, along with the immaturity of the design of large main battery guns, had limited their effective firing range with the weapons typically being fired over open sights. Lacking any form of centralized fire-control system also meant the various turreted guns on battleships could not be directed at the same target as each turret gun crew individually aimed their weapons at the target or targets they felt were the most important. In addition, the impact splashes from the secondary and intermediate gun batteries often confused the spotters for the main battery guns.

The Best and the Brightest Take up the Cause

The push to replace 'mixed gun' battleships with 'all-big-gun' battleships equipped with then state-of-the-art fire-control systems in the US navy was driven by a small number of innovative officers. The first of these influential young men was Bradley A. Fiske, who later rose to the rank of rear admiral. His time in service spanned from 1874 to 1916. Early in his career as a lieutenant, he identified the need for American ship guns to be equipped with telescopic range-finders to improve accuracy. He then set about designing and patenting suitable devices for use on US navy ships. He later invented and patented a turret range-finder, an important development since finding the correct range to a target is of critical importance in accuracy. Taken together, these inventions laid the groundwork for the fire-control systems found on all large US navy ships from 1908 onwards.

Another bright young man who did much to drag the US navy into accepting all-big-gun battleships was William S. Sims, who later rose to the rank of vice admiral. He served in the US navy from 1880 until 1922. On assignment in the Far East, Sims had the good fortune to meet Captain Sir Percy Scott of the Royal Navy who was then serving in the Royal Navy's Asiatic Squadron. Scott was a gunnery expert who introduced Sims to the latest methods of improving gunfire accuracy for ships. Sims embraced Scott's methods but when he was rebuffed by the US navy, he wrote to President Theodore Roosevelt in November 1901. Roosevelt eventually took up his cause and forced the US navy to improve its fire-control and gunnery practices.

Sims also helped to convince Roosevelt of the merits of all-big-gun battleships. He was aided in this endeavour by Lieutenant Commander Homer C. Poundstone of the US navy, who created blueprints of a ship he named the *Possible*, armed with twelve 11-inch main battery guns.

In June 1903 naval architect US navy Captain Washington Irving Chambers built upon Poundstone's idea for an all-big-gun battleship and submitted it to the Naval War College, which war-gamed the concept and found it had merit. This resulted in a feasibility study in October 1903 that proposed a battleship armed with twelve 12-inch main battery guns, no intermediate battery guns, and a secondary battery of pedestal-mounted 3-inch guns aimed at dealing with torpedo-boat attacks. It was this proposed all-big-gun battleship concept that eventually became the foundation of the US navy's two-battleship *South Carolina* class.

In spite of being at the forefront of ship design in 1903 with the concept of an all-big-gun battleship, the US navy's efforts at having such ships authorized by Congress found themselves dashed on the shoals of a byzantine bureaucracy. It took until March 1905 before Congress authorized the building of the two *South Carolina*-class battleships. They were not commissioned until 1910 and in the meantime, a number of other navies grasped the advantages of switching to all-big-gun battleships. In 1904 the Japanese navy began construction of an all-big-gun battleship, named the *Satsuma*, to be armed with twelve 12-inch guns. However, their plans were not realized due to a shortage of 12-inch guns and the ship was eventually built as a mixed-gun battery battleship.

A New Type of Battleship

It was the Royal Navy that would take credit for launching the first all-big-gun battleship, named HMS *Dreadnought*. It was designed in early 1905 and launched in early 1906. The prefix 'HMS' stands for 'His (or Her) Majesty's Ship'. HMS *Dreadnought* had five main gun battery turrets including two wing turrets, each armed with two 12-inch guns. There were no intermediate gun batteries. However, there were twenty-seven smaller 12-pounder (3-inch) guns making up the ship's secondary battery, mounted on the roof of the main gun turrets and around the ship's superstructure.

With a full load displacement of approximately 18,000 tons, HMS *Dreadnought* was also the first battleship with steam turbine engines powered by coal-fired boilers. This was in contrast to the triple-expansion reciprocating steam engines powered by coal-fired boilers found in all the world's battleships until that time. The steam turbine engines gave HMS *Dreadnought* a top speed of 21 knots, making it the fastest ship of its type in the world in its time. Steam turbine engines were capable of generating higher ship speeds due to their greater thermal efficiency, providing a higher power-to-weight ratio. Those engines could also keep up their maximum speed for much longer than triple-expansion reciprocating steam engines, which was even more important than the top speed. Steam turbine engines also took up less space and were more durable than the existing triple-expansion steam engines.

The superior firepower and speed of HMS *Dreadnought* made every battleship in service obsolete overnight. It soon caused a frantic arms race among the other major navies of the world to maintain parity in strength with the now more advanced Royal Navy by building their own versions of the new ship design. As HMS *Dreadnought* became the new benchmark by which all battleships were measured, battleships commissioned before it became known as 'pre-dreadnought battleships'. The all-big-gun battleships that followed the commissioning of HMS *Dreadnought* are generally referred to as 'dreadnoughts', although that term fell out of favour after the First World War.

South Carolina Class

The first of the all-big-gun dreadnought battleships for the US navy were the *South Carolina* class, consisting of the USS *South Carolina* (BB-26) and USS *Michigan* (BB-27). They were authorized by Congress in March 1905 and commissioned in 1910.

The primary armament of the *South Carolina*-class battleships consisted of eight 12-inch main battery guns mounted in four large turrets on the centreline of the ship, each of which was armed with two guns. Two of the main gun battery turrets were mounted forward of the ships' superstructure and two aft. In a more efficient layout than present on HMS *Dreadnought*, there were no main gun battery wing turrets on the *South Carolina*-class battleships.

The main battery gun turrets on the *South Carolina*-class ships were superimposed on each other, with the lower main battery turret fitted on the weather/protective deck on a barbette. The second main gun battery turret was mounted just behind and above the lower turret on a taller barbette that cleared the lower main battery turret. This design arrangement was known as 'super-firing' turrets. A 'barbette' is a heavily-armoured cylinder extending from the upper part of a gun turret down to the handling and magazine areas of a ship.

The *South Carolina*-class battleships had, besides their main gun batteries, an array of smaller pedestal-mounted QF secondary guns for close-in protection from torpedo boats that were either casemated or mounted on various decks or levels of the ships.

The sloping vertical faces of the main gun battery turrets on the *South Carolina*-class battleships were protected by 12 inches of armour. The hull belt armour had a maximum thickness of 12 inches, while the protective deck on the ships had a maximum thickness of 2 inches of armour. All the ships had double bottoms.

Prior to the design and building of the *South Carolina*-class battleships the US navy had become more aware of the threat posed by underwater mines and torpedoes. As a result, the ships had their coal bunkers located adjacent to the bottom portion of their hulls. These bunkers were separated by bulkheads (traverse or longitudinal partitions separating portions of a ship) intended to act as a cushion and absorb and dissipate the detonation of an underwater mine or torpedo. This was a design feature adopted on subsequent US navy battleship classes.

The USS *South Carolina* (BB-26) and the USS *Michigan* (BB-27) were both smaller and lighter than HMS *Dreadnought* with a full load displacement of 17,617 tons and a crew complement of 869 officers and enlisted men. This figure rose to 1,354 officers and enlisted men during wartime. The length of the ships was 452 feet 9 inches, with a beam of 80 feet 3 inches.

Unlike the steam turbine engines of HMS *Dreadnought*, the *South Carolina*-class battleships retained the triple-expansion steam engines powered by coal-fired boilers which drove all the US navy pre-dreadnought battleship classes that came before them. The decision to stick with the older generation propulsion system on US navy battleships reflected their greater fuel efficiency over the existing steam turbine engines. Due to the greater distances that US navy ships had to operate in the Pacific Ocean compared to their British and European counterparts, fuel efficiency was seen as a more important attribute than speed. The *South Carolina*-class battleships had a maximum speed of 18.5 knots.

The *South Carolina* class was the first to feature two very tall cage masts (also known as lattice masts) that became a standard feature for all American dreadnoughts for many years. The cage masts were intended to provide tall, stable platforms for fire-control devices and were retro-fitted to the US navy pre-dreadnought battleships when they returned to shipyards for refit and overhaul. They replaced the former military masts on US navy battleships and were felt to be better able to resist enemy shellfire by allowing blast to pass through them. The *South Carolina*-class battleships saw their final decommissioning by 1922.

Delaware Class

The two *South Carolina*-class battleships were followed into service by the two *Delaware*-class battleships consisting of the USS *Delaware* (BB-28) and USS *North Dakota* (BB-29). These American battleships were nearly a match in size and capabilities to HMS *Dreadnought*, as Congress had finally removed its previous 16,000-ton restriction on the tonnage available to the US navy when designing its battleships.

The main battery gun arrangement of the *Delaware*-class battleships was centreline on the ship and consisted of five two-gun turrets armed with 12-inch guns, two of them located in front of the ships' superstructures and the other three aft. Of the three 12-inch gun turrets located aft of the superstructure, the one located directly behind it was elevated on a barbette and could fire over the two 12-inch gun turrets located behind it on the weather/protective deck. The two 12-inch gun turrets mounted on the rear weather/protective deck of the *Delaware*-class battleships were located back-to-back. They were therefore limited to firing over the sides of the ship only. This was not an issue at the time as the US navy, like every other navy, expected to fight a big-gun duel in opposing columns firing broadsides at each other. Secondary armament on the *Delaware*-class ships was composed of fourteen pedestal-mounted 5-inch guns either casemated or mounted on the various decks or levels of the vessels.

Armour thickness on the sloping vertical faces of the main gun turrets on the *Delaware*-class battleships was 12 inches. Their hull belt armour had a maximum thickness of 11 inches and the protective deck 3 inches. Their conning towers were built from armour 11.5 inches thick. Like the *South Carolina*-class battleships they had double bottoms and had their coal bunkers moved to the bottom of their hulls for added stand-off protection from underwater mines or torpedoes.

The USS *Delaware* (BB-28) retained the triple-expansion reciprocating steam engines of its predecessors but was now capable of a top speed of 21 knots, as was its Royal Navy counterpart. Its boilers were coal-fired.

The USS *North Dakota* (BB-29) was the first battleship in the US navy with steam turbine engines. In another first, the ship employed both coal-fired and fuel oil-fired boilers to power its engines and this provided the ship with a maximum speed of 21 knots, now the accepted norm for all dreadnought-inspired battleships. The reason for having coal and fuel oil boilers on the *North Dakota* was the poor fuel consumption rates with early fuel oil-fired boilers at the time.

The USS *Delaware* was authorized in 1906 and her sister ship the following year, with both ships commissioned in 1910. Both battleships had a full load displacement of 22,000 tons and a peacetime crew of 933 that was raised to 1,384 during times of war. The length of the ships was 518 feet 10 inches with a beam of 85 feet 3 inches.

Florida Class

Following in the footsteps of the *Delaware*-class battleships that were both decommissioned in 1923 came the two *Florida*-class battleships, the USS *Florida* (BB-30) and the USS *Utah* (BB-31), authorized by Congress in 1908. Both ships were slightly enlarged versions of the *Delaware*-class battleships and in contrast to their predecessors had four propeller shafts instead of two. The ships had steam turbine engines originally powered by coal-fired boilers but later converted to fuel oil-fired boilers which gave them a top speed of 21 knots.

The two *Florida*-class battleships were commissioned in 1911 and had their final decommissioning in 1931. With a full load displacement of 23,033 tons, the ships were crewed by 1,001 officers and men in peacetime and 1,492 during wartime. They had a length of between 521 feet 6 inches and 526 feet 6 inches, with a beam of 88 feet 3 inches.

Originally intended to be armed with eight 14-inch main battery guns divided between four two-gun centreline armed turrets, a delay in their production and testing meant the two *Florida*-class ships received the same ten 12-inch main battery guns, divided between five centreline turrets, that had been fitted to the *Delaware* class. The *Florida* class secondary battery guns armament consisted of sixteen casemated 5-inch guns.

Armour on the sloping vertical faces of the main gun batteries on the *Florida*-class battleships was 12 inches thick. The hull belt armour on the ships had a maximum thickness of 11 inches, with their protective decks having a maximum thickness of 1.5 inches. Conning tower armour was 11.5 inches thick.

Both *Florida*-class battleships went through a modernization process in the 1920s that increased their armour protection below the waterline. This was accomplished by adding external torpedo bulges, also known as 'blisters', to either side of their lower hulls. This had been a British invention and was added to many of their First World War ships. These external bulges were intended to absorb and dissipate the blast from underwater mine or torpedo strike, thus preventing damage to the ship's actual double hull located behind the external bulges.

Additional improvements to the *Florida*-class battleships during the 1920s modernization process included an aircraft-launching catapult on one of their aft main gun battery turrets. Prior to that point, some of the previous battleship classes had been fitted with short flying-off platforms mounted on the roofs of both forward and aft main gun battery turrets. The aircraft employed from these platforms were small and lightweight First World War vintage designs and were intended as observation planes. As aircraft designs progressed in the interwar years and planes became larger and heavier, these short flying-off platforms were no longer adequate and were replaced by catapults.

Wyoming Class

Following the *Florida*-class battleships came the two even larger *Wyoming*-class battleships: the USS *Wyoming* (BB-32) and the USS *Arkansas* (BB-33). They were authorized in 1910 and commissioned in 1912. These ships were based on the *Florida*-class designs but were a bit larger, with a full load displacement of 27,243 tons. They had a crew complement of 1,063 officers and men in peacetime and 1,594 in wartime. Length was 562 feet, with a beam of 93 feet 3 inches.

There was some thought of mounting centreline 14-inch main gun battery guns on the *Wyoming*-class battleships as had been considered for the *Florida* class but this did not happen and they were armed with 12-inch centreline main battery guns. To increase their firepower, an added 12-inch gun-armed main battery gun turret was mounted aft of the ships' superstructure, bringing the total to six. This meant there were now four 12-inch gun-armed main gun turrets at the rear of the battleships: two mounted on tall projecting barbettes and two on barbettes flush with the weather/protective decks of the ships.

The armour thickness on the *Wyoming*-class battleships was the same as on the preceding *Florida* class. As with their predecessor class battleships, the *Wyoming*-class ships went through a modernization process in the 1920s that added torpedo bulges to their double-bottom hulls.

The casemated hull-mounted secondary gun batteries on the *Wyoming*-class battleships proved unusable due to being almost constantly inundated by seawater, particularly those at the front and rear of the hull, and were later plated over. Some of these secondary gun batteries were later moved to casemated positions on the battleship's upper decks (partial decks above the weather/protective deck amidships). This arrangement was also seen on the dreadnought-type battleships that followed them into service with the US navy.

Like the *Florida*-class battleships, the *Wyoming* class had steam turbine engines that were originally powered by coal-fired boilers and then converted to fuel oil-fired boilers. Top speed of the ships was 21 knots.

As with other early dreadnought-inspired vessels, the *Wyoming*-class battleships went through a number of upgrades over the years to maintain their combat capabilities against foreign battleships. Both battleships saw service during the Second World War with the USS *Wyoming* only being employed as a training ship, whereas the USS *Arkansas* saw combat in both the European and Pacific theatres of operations.

Both ships had their final decommissioning shortly after the conclusion of the Second World War.

The USS *Arkansas* was one of four US navy battleships employed as target ships during Operation CROSSROADS; the testing of two atomic bombs at Bikini Atoll in the Pacific Ocean in July 1946. The battleship sank upon detonation of the second atomic bomb on 25 July 1946.

New York Class

It was the two *New York*-class battleships, authorized in 1910 and commissioned in 1914, that finally received the 14-inch main gun batteries that the US navy had desired since the *Florida* class. Reflecting the increased firepower and size of the weapons, the USS *New York* (BB-34) and the USS *Texas* (BB-35) had only five centre-line main gun battery turrets, each armed with two 14-inch guns. The turret configuration reverted to that seen on the *Delaware*- and *Florida*-class battleships, with two main gun battery turrets located forward of the ships' superstructures and three aft.

Secondary battery armament originally consisted of twenty-one 5-inch casemated guns on the *New York*-class battleships. This was upgraded with a large number of small-calibre anti-aircraft guns before and during the Second World War.

Armour on the sloping vertical face of the main gun batteries of the *New York*-class ships was 14 inches thick. The vertical hull belt armour had a maximum thickness of 12 inches, with the protective deck having a maximum thickness of 2 inches. Conning tower armour was 12 inches thick.

With the *New York*-class battleships the US navy reverted to triple-expansion steam engines powered by coal-fired boilers. All subsequent US navy battleships classes were designed from the beginning to have fuel oil-fired boilers. Both *New York*-class ships were later modernized to have fuel oil-fired boilers. The US navy's decision to use triple-expansion reciprocating steam engines on the *New York*-class battleships reflected their greater fuel economy compared to the steam turbine engines of the day.

The two *New York*-class battleships had a crew complement of 1,042 officers and enlisted men in peacetime and 1,612 in wartime. Both ships had an original full load displacement of 28,367 tons, although that of the USS *Texas* was later raised to 32,000 tons after going through a modernization process that was completed by 1927. Both *New York*-class battleships had an overall length of 573 feet with a beam of 93 feet 3 inches. As with other dreadnoughts that went through a modernization process in the 1920s and 1930s, they lost their submerged torpedo tubes.

A key feature of the 1920s modernization process for the USS *Texas* and other dreadnought-type US navy battleships was the addition of external torpedo bulges. As the beam of all capital ships of the US navy could not exceed 110 feet in order to allow them to pass through the locks of the Panama Canal (opened in 1914), the torpedo bulges on the ship added only 5 feet on either side of their beam. Additional features of the modernization process were better fire-control systems, increased elevation of the main battery guns and added horizontal protection.

Reflecting their upgrading, both *New York*-class battleships saw service in the Second World War and went on to have their final decommissioning when the conflict concluded. The USS *New York* was one of four US navy battleships used as target ships during Operation CROSSROADS. It survived both atomic bomb tests conducted at Bikini Atoll in July 1946 and was later sunk off the Hawaiian coast during additional testing by the US navy in July 1948.

The USS *Texas* had its final decommissioning in 1948. Since that time the *Texas* has had a second life as a museum ship located at San Jacinto, Texas as the only surviving dreadnought battleship of the US navy, and in the world. No pre-dreadnought US navy battleships have survived till the present day.

The USS *South Carolina* (BB-26) was built by William Cramp & Sons and was the first all-big main gun battery dreadnought-type battleship in the US navy. Its career, like many of its predecessor pre-dreadnought battleships, proved uneventful. During the First World War it patrolled along the East Coast of the United States and later served in the convoy protection role and as a troop transport. (*National Archives*)

Following the First World War, the USS *South Carolina* spent time taking midshipmen from the US Navy Academy on various educational cruises in both the Atlantic and Pacific oceans. The US navy decommissioned the battleship in December 1921. It was struck from the navy list in November 1923 and scrapped as shown here in April 1924, in accordance with an arms control treaty agreed by the United States government. (*National Archives*)

The second ship comprising the *South Carolina* battleship class was the USS *Michigan* (BB-27). It was built by the New York Shipbuilding Corporation, New Jersey. Its forward cage mast collapsed in January 1918 during heavy seas, resulting in the deaths of six sailors with another thirteen injured. Like its sister battleship the USS *South Carolina*, the USS *Michigan* never fired a shot in anger and was eventually scrapped in 1924. (*National Archives*)

A close-up picture of the aft main gun battery turrets on the USS *Michigan* with numerous secondary battery guns seen on the ship's superstructure, covered with canvas tarps for protection from the elements. Firing armour-piercing rounds, the eight 12-inch main battery guns had a maximum range of 15,000 yards or roughly 8.5 miles, although the chance of hitting at this distance was very small given the crude gunnery techniques of the day. (*Library of Congress*)

The painted lines on the main gun battery turrets, seen here on the USS *Michigan*, were referred to as 'bearing (deflection) markings'. Employed in conjunction with the range clock seen on the forward cage mast, the bearing markings on the main gun battery turrets and the range clock allowed other US navy battleships in close proximity to determine the azimuth (direction) and elevation for aiming their own main guns when their targets might be obscured by smoke. This was done in an era before reliable ship-to-ship radio communication. (*National Archives*)

Shown at its fitting-out dock is the USS *Delaware* (BB-28), the first battleship of the two-ship *Delaware* class. It was constructed by the Newport News Shipbuilding Company and commissioned in 1910. Visible on the forward portion of the port side of the hull is an armoured sponson mounting a 5-inch secondary gun that was duplicated on the starboard side. Both of these armoured sponsons and their guns were removed by 1912, as is evident from dated photographs of the battleship. (*Maritime Quest*)

A stern view of the USS *Delaware* moored in a warm-weather anchorage, as is evident from the large number of sailors swimming in the water around the ship. Such liberties were good for the morale of the crew who lived a very stern and regimented existence on their ships. Visible on the roof of each of the ship's three aft main gun battery turrets are armoured stereoscopic range-finders, referred to as the Mk 31 director, installed after the First World War.
(*National Archives*)

The USS *Delaware* spent time in service working alongside the Royal Navy's Grand Fleet in the First World War and later performed convoy protection duties. The ship's only combat encounter during the First World War occurred off the waters of Stavanger, Norway, when it successfully evaded two torpedoes fired by a German submarine. This photograph shows the ship on 1 January 1920 off the coast of Cuba. The vessel was sold for scrapping in 1924.
(*National Archives*)

Prior to the mounting of armoured range-finders (referred to as directors) on the main gun battery turrets of US navy dreadnought-type battleships, there was experimental mounting of open stereoscopic range-finders on some ships to test their effectiveness, as seen here. For effective fire control of a ship's main battery guns, one of the most important pieces of information was the range to a target. (*National Archives*)

Constructed by the Fore River Company, Massachusetts was the USS *North Dakota* (BB-29) which never fired a shot in anger during its career. However, it played a supporting role in the landing of American Marines and armed sailors at Vera Cruz, Mexico in April 1914. The landing resulted in some initial fighting and then a six-month occupation of the city by the American military. The excuse for the landing was the jailing of some American sailors on what were perceived to be bogus charges. (*National Archives*)

During the First World War the USS *North Dakota*, unlike her sister battleship the USS *Delaware*, did not serve overseas. It remained in American waters throughout the conflict as a training ship and was decommissioned in 1923 and disarmed. From 1924 through 1930 it was employed as a radio-controlled target ship until replaced by another retired battleship in that role. The *North Dakota* was sold for scrapping in 1931. (*National Archives*)

The USS *Florida* (BB-30) was built at the New York Navy Yard, Brooklyn. It was launched as seen in this photograph on 12 May 1910 and towed to a fitting-out dock where the engines and boilers, as well as the ship's superstructure and weapons, were added. The battleship lasted in US navy service until it was decommissioned in 1931 and then scrapped in accordance with an arms control treaty. (*National Archives*)

This overhead photograph shows the open navigation bridge and bridge wings of the USS *Florida*. The ship spent a period of time serving alongside the Royal Navy's Grand Fleet during the First World War. The battleship was also present at Scapa Flow, Scotland, when the Imperial German High Seas Fleet sailed in to surrender on 21 November 1918. (*National Archives*)

A picture taken on the fantail (the aft end of the weather deck) of the USS *Florida* shows the large number of searchlights mounted on the ship, the majority being located on the cage masts. Prior to the fitting of radar on US navy battleships starting shortly before the Second World War, one of the few ways that such battleships could engage in combat at night was with the aid of searchlights. (*National Archives*)

A stern view of a modernized USS *Florida* that had its aft cage mast removed and replaced with a simple pole mast. Another visual clue to the modernization of the battleship that took from April 1925 until November 1926 was the deletion of one of the ship's two stacks. This reflected the fact that the battleship's numerous coal-fired boilers had been replaced by a lesser number of more efficient fuel oil-fired boilers that emitted less exhaust. (*National Archives*)

Visible in this picture of the modernized USS *Florida* is a fixed armoured range-finder/director on top of the roof of the forward super-firing main gun battery turret. There is also a traversable armoured range-finder/director on the roof of the navigation bridge, which is now enclosed. Above the navigation bridge is a flag (admiral's) bridge. The main observation post for the ship, and for controlling the fire of the main battery guns, was located at the top of the cage mast. (*National Archives*)

The navigation bridge of a US navy battleship, seen here, has its armoured shutters (fitted with vision slots) closed as would be the case prior to battle. The navigation bridge was the typical station for the officer controlling a ship during the early part of the twentieth century. Also visible in this picture is the large wheel for steering the battleship. Other devices present include the binnacle, a stand that housed the magnetic compass, and the engine order telegraph, or annunciator, by which the orders of the conning officer would be passed to the engine room. (*National Archives*)

This aerial view of the modernized USS *Florida* shows the centreline setting of the three aft main gun battery turrets of the ship. This arrangement of main gun battery turrets first appeared on the *Delaware*-class battleships. Mounted on the top of the aft main gun battery turret No. 3 is a US navy observation floatplane, set on a launching catapult. It observed and reported the fall of the ship's rounds so that more accurate fire-control solutions could be made. At one time the US navy experimented with using manned observation balloons in that role, without success. (*National Archives*)

The USS *Utah* (BB-31), shown here, was the sister ship of the USS *Florida*. It was built at the New York Shipbuilding Corporation, New Jersey and is pictured here in a First World War camouflage paint scheme. During that conflict the battleship, alongside other US navy battleships, was tasked with protecting allied convoys from Imperial German Navy ships. (*National Archives*)

A stern view of the USS *Utah* displays its three aft main gun battery turrets, labelled turret Nos 3, 4 and 5 from front to back. At the conclusion of the First World War the battleship performed a number of missions. These included escorting the ocean liner that transported American President Woodrow Wilson to France for the postwar peace negotiations that led to the Treaty of Versailles. The treaty was signed on 28 June 1919 and set strong restrictions on the rebuilding of the German military, including ground, aerial and naval forces. (*Maritime Quest*)

The USS *Utah* pictured here took the commander of American troops in France during the First World War, now General of the Armies John J. Pershing, on a goodwill tour of South America that lasted from late 1924 until March 1925. Between October and December 1925 the ship was modernized, as were other US navy battleships of that era, including the replacement of its coal-fired boilers with fuel oil-fired boilers and the aft cage mast replaced by a pole mast. (*National Archives*)

In 1931 the USS *Utah*, shown here off the California coast, was de-gunned and reclassified as a radio-controlled target ship. The former battleship was employed to train anti-aircraft gunners from 1935 onwards. The vessel was present at Pearl Harbor when the Japanese attack occurred on 7 December 1941, was struck by two torpedoes and sank at its mooring station with the loss of sixty-four men. It remains at Pearl Harbor to this day as a memorial. (*National Archives*)

The USS *Wyoming* (BB-32), pictured here running at high speed, was constructed by William Cramp & Sons. During the First World War it spent time working alongside the Royal Navy's Grand Fleet and was present when the Imperial German Navy High Seas Fleet surrendered in November 1918. The battleship then returned to the United States in December 1918. (*National Archives*)

A stern view of the USS *Wyoming* passing through the Panama Canal in July 1919. This photograph clearly shows the four centreline main gun battery turrets mounted aft of the ship's superstructure, a feature not seen on any US navy battleship class before or after it. Some thought had been given to mounting 14-inch main battery guns on the *Wyoming*-class battleship but the US navy decided to stick with the 12-inch main battery guns to speed up its commissioning. (*National Archives*)

The USS *Wyoming* was modernized in 1927, losing one of two stacks, as is seen in this picture. To improve its underwater protection from torpedo strikes, external watertight compartments referred to as 'bulges' or 'blisters' were added to the ship's lower hull, as was done with many other US navy dreadnought-type battleships. Located above the ship's navigation bridge is a flag (admiral's) bridge. (*National Archives*)

Common to the US navy battleships that were modernized in the 1920s, the USS *Wyoming* had its rear cage mast replaced by a sturdier tripod mast, as seen in this photograph. On the roof of its aft main gun turret No. 3 is a launching catapult with two observation floatplanes stored on it. The ship underwent a demilitarization process to meet arms control treaty obligations in 1931, which included the removal of two of her main gun battery turrets. (*National Archives*)

Following America's official entry into the Second World War, there were thoughts of returning the USS *Wyoming* to full battleship status. However, that did not happen and the ship was employed instead to train anti-aircraft gunners. Eventually the ship's remaining main gun battery turrets were removed as seen in this picture and replaced with smaller turrets armed with twin and single 5-inch dual-purpose guns. The vessel was sold for scrapping in 1947. (*National Archives*)

The sister ship of the USS *Wyoming* was the USS *Arkansas* (BB-33), shown here in dry dock at the New York Shipbuilding Corporation, New Jersey, the builders of the battleship. Like some of the other US navy dreadnought-inspired battleships, the USS *Arkansas* spent time serving alongside the Royal Navy's Grand Fleet during the First World War and was present when the Imperial German Navy High Seas Fleet surrendered in November 1918. (*National Archives*)

Shown here at the end of the First World War is the USS *Arkansas*. Between 1919 and 1921 the battleship served with the US navy's Pacific Fleet. Between 1925 and 1926 the battleship was modernized which included the replacement of its coal-fired boilers with fuel oil-fired boilers and the aft cage mast being replaced with a tripod mast. From December 1941 until September 1945 the ship would see action in both the Atlantic and Pacific Ocean theatres of operation. The vessel ended its career as a target ship for atomic bomb tests and its hulk still rests on the bottom of Bikini Atoll. (*National Archives*)

The USS *New York* (BB-34) was constructed at the New York Navy Yard, Brooklyn, as shown here. As with other dreadnought-inspired US navy battleships, it spent time reinforcing the Royal Navy's Grand Fleet during the First World War. It fired no shots in anger during that conflict but accidentally ran into and sank a German navy submarine that had the misfortune to be under the battleship as it passed overhead. (*Maritime Quest*)

The USS *New York* was modernized in the 1920s. The key external feature of the modernization process was the replacement of the original cage masts with the new tripod masts seen here, as happened with other dreadnought battleships in US navy service. The forward tripod mast was in naval terms known as the 'foremast' and the rear tripod mast was referred to as the 'main mast'. (*National Archives*)

A stern view of the USS *New York* taking part in some type of official ceremony as is evident from the crew lining the rails. During the Second World War the battleship served in the role of convoy protection and also provided gunfire support during Operation TORCH, the Allied invasion of French-held North Africa in November 1942. (*National Archives*)

In late 1938 the USS *New York* was fitted with an experimental radar system, designated XAF, seen here mounted on the roof of its navigation bridge. It was the second ship in the US navy to have a radar system fitted. The first experimental radar system had been installed on the destroyer USS *Leary* the previous year. The radar system was designed to detect aircraft at ranges up to 50 miles and large ships at 14 miles. (*National Archives*)

The USS *New York* was chosen for the role of a gunnery training ship, beginning in 1943. This role continued until late 1944 when the battleship was sent to the West Coast of the United States for refresher training before being deployed to the Pacific theatre of operations in January 1945. In conjunction with other older-generation US navy battleships, it provided gunfire support during the invasion of Iwo Jima in February 1945. (*National Archives*)

The USS *New York* was being refitted at an American shipyard when the Japanese attack took place at Pearl Harbor on 7 December 1941. After returning to service it was tasked with escorting Allied convoys in the Atlantic until June 1942. The ship then went into another refit to have the number of small-calibre anti-aircraft guns increased, as is reflected in this wartime picture of the battleship. (*National Archives*)

From the forecastle of the USS *New York* looking rearward we can see the two forward main gun battery turrets of the battleship, turrets Nos 1 and 2, as well as the forward tripod mast, during a refit. The radar antenna seen on the roof of the main battery fire control, mounted on top of the tripod mast, identifies this photograph as being taken during the Second World War. Note the addition of numerous circular, open-topped gun tubs on the navigation bridge, as well as the foremast for mounting small-calibre anti-aircraft guns. (*Maritime Quest*)

Reflecting the age of the USS *New York*, it did not last long in US navy service following the conclusion of the Second World War. It was chosen to serve as one of numerous target ships during the testing of atomic bombs in the Pacific in 1946. The old battleship survived the detonation of two atomic bombs, only to be employed for target practice off the coast of Hawaii as seen in this July 1948 picture. The ship eventually succumbed to its numerous wounds after this picture was taken and sank to the bottom of the ocean. (*National Archives*)

A picture of the USS *Texas*'s (BB-35) nearly completed hull prior to its launching ceremony which took place on 18 May 1912. US navy ships traditionally are not given the prefix 'USS' until commissioned and officially accepted into service. The two circular openings on the lower starboard side were for the ship's underwater torpedo tubes, with another two being present on the port side. (*Library of Congress*)

This 1914 photograph of the USS *Texas* shows the ship's open navigation bridge; a common feature in its day, covered with a canvas windscreen for protection from the elements. Just below the navigation bridge and located behind the rear super-firing main gun battery turret is the armoured conning tower. (*Library of Congress*)

Heading towards the Brooklyn Bridge in New York Harbor in 1914 is the USS *Texas*. This view of the ship shows the arrangement of the three aft main gun battery turrets. Also visible are some of the casemated 5-inch guns located just below the battleship's weather deck, including the one that was located at the stern. (*Texas SHS*)

This photograph of the USS *Texas* shows its appearance sometime between 1916 and 1917. During the night of 27 September 1917 the ship ran aground off Block Island, Rhode Island, suffering serious damage to its lower hull. The battleship was then escorted by tugs to New York Harbor and went into dry dock at the New York Navy Yard for repairs that took from October to November 1917. (*Library of Congress*)

The navigation bridge on the USS *Texas* was enclosed as seen here in this 1918 photo. Above the navigation bridge is a stereoscopic range-finder. Mounted on the front of the cage mast just above and behind the range-finder is a large clock-like signalling device, referred to as the 'range dial'. It displayed to other US navy battleships the range to a target that might be obscured from their own observation. (*National Archives*)

Passing through the locks of the Panama Canal is the USS *Texas*. On the roof of main gun battery turret No. 2 is a 1920s-era biplane. It was intended as an observation plane for the ship. The *Texas* was the first US navy battleship to successfully launch an aircraft from a flying-off platform on 9 March 1919. Other US navy dreadnought-type battleships were eventually fitted with the same arrangement. (*National Archives*)

A close-up picture of a First World War-era Sopwith Camel employed as an observation plane mounted on the roof of the super-firing forward main gun battery turret No. 2 of the USS *Texas*. The aircraft is in its stored position with the wings clamped to the brackets on either side of the main gun battery turret roof and the aircraft's engine is covered

with foul-weather gear. Visible directly behind the aircraft is the ship's armoured conning tower and behind and above that, the enclosed navigation bridge.

(*National Archives*)

At some point it was decided to erect a second launching platform for an observation biplane on the USS *Texas*. This was placed on the roof of the aft main gun battery turret No. 3, as seen here. Visible is the metal framework attached to the two barrels of the main gun battery turret. Prior to launching the aircraft, the metal framework was covered with wooden planks. (*National Archives*)

The USS *Texas* is seen here in dry dock sometime after the conclusion of the First World War. The port torpedo-tube launching openings are seen on the bottom portion of the hull. The firing positions for the forward casemated 5-inch guns are plated over in this photograph as they proved useless in anything other than calm weather conditions. Visible above the enclosed navigation bridge is an enclosed flag (admiral's) bridge. (*National Archives*)

Sailors of the USS *Texas* pose for the photographer on the ship's aft main gun turrets, Nos 4 and 5. Both of the *New York*-class battleships, the USS *New York* (BB-34) and the USS *Texas* (BB-35), were the first of their type in the US navy to be fitted with 14-inch main battery guns. They had a range, when firing armour-piercing projectiles, of 18,000 yards or just over 10 miles. (*National Archives*)

Shown on the weather deck of the USS *Texas* is a recovered Whitehead Mk 5 torpedo that had been employed in a practice firing exercise. The torpedo had been developed in 1901 and was in service on US navy battleships from about 1910 until the mid-1920s. The *Texas* had its underwater torpedo tubes removed, as did many other US navy dreadnought-type battleships, during a refit and modernization that took place from the mid-1920s. (*Texas SHS*)

Pictured is the No. 4 main gun battery turret of the USS *Texas*. It has two 3-inch anti-aircraft guns mounted on its roof. As with all battleships, it was the main gun battery that provided the reason for its existence. The US navy's decision to mount 14-inch main gun battery armed turrets on the *New York*-class battleships was driven by news that the Royal Navy was going to introduce into service a 13.5-inch gun for its newer battleship main gun turrets. The cylinders on the ship's weather deck held the silk bags that contained the propellant for the main guns. (*Texas SHS*)

This photograph shows the USS *Texas* between its modernization that took place between 1925 and 1926 and the addition of the enclosed flag (admiral's) bridge around 1927. Gone are the cage masts, replaced by tripod masts. In addition, the ship's fourteen original coal-fired boilers were replaced by fuel oil-fired boilers and one of the two stacks was dispensed with. Directly aft of the now single stack is a new fire-control tower. (*National Archives*)

A close-up picture of the launching catapult fitted to the roof of aft main gun battery turret No. 3 on the USS *Texas* following its 1925–26 modernization. Stored on the launching catapult are three observation floatplanes. The catapult launched the 6,000-lb observation floatplane at a speed of 70mph. In front of the forward-facing observation floatplane is the new aft fire-control tower. (*Texas SHS*)

The USS *Texas* (BB-35) is the only US navy dreadnought-type battleship to survive till the present day. The elderly ship has been a museum ship since 1948. It is seen here moored in a slip alongside the Houston Ship Channel. Due to its age and the damage inflicted upon the battleship's hull by decades of corrosion, current plans call for the ship to be moved on land for its long-time survival, if funding can be found. *(Paul Hannah)*

This contemporary picture shows one of the six 5-inch secondary guns that the USS *Texas* had at the conclusion of the Second World War. When commissioned in 1914 the battleship had twenty-one 5-inch secondary guns fitted. The weapon type itself first appeared on the USS *Florida* (BB-30) commissioned in 1911 and remained the standard secondary gun on US navy battleships commissioned up until 1923. The weapon was suitable only for use against surface targets, lacking the elevation to engage aerial targets. *(Paul Hannah)*

On the main deck of the USS *New Mexico* (BB-40), a super dreadnought-type US navy battleship, is a battery of four 5"/25 dual-purpose secondary battery guns. The photograph dates from the Second World War and is one of a very small number taken in colour of US navy battleships during that conflict. The weapon fired fixed (one-piece) ammunition that was set to explode at a pre-determined altitude. This was accomplished by placing the fuses of the upturned rounds into the fuse-setter located at the rear left-hand side of the gun mount, as seen in this picture.

(*National Archives*)

Taken on board an *Iowa*-class battleship is this picture of a young US navy sailor inserting a 5-inch projectile into a power hoist that will take it up to the ammunition handling room which is located directly below a 5"/38 dual-purpose secondary battery gun. From the ammunition handling room, sailors will insert the projectile into another power hoist that will bring it up into the gun mount. The 5"/38 dual-purpose secondary battery guns fired semi-fixed ammunition in which the primer and propelling charge are firmly fixed in the cartridge case but the projectile is separate. (*US Navy*)

Shown in dry dock is the USS *Iowa* (BB-61) that had been modernized in the 1980s, with new electronics and weapons. The circular splinter shield seen at the very tip of the battleship's bow had contained two 20mm anti-aircraft guns during the Second World War. They were removed prior to the ship's service in the Korean War but the splinter shield was retained. This is a feature seen on another *Iowa*-class battleship, the USS *Wisconsin* (BB-64) but not the other two ships of this class; the USS *New Jersey* (BB-62) and the USS *Missouri* (BB-63). (*US Navy*)

A main gun battery salvo is fired from a modernized *Iowa*-class battleship. A salvo is defined by the US navy as the firing of two or more guns simultaneously on the same signal at the same target. Salvos are normally fired at regular intervals that are spaced to permit reloading and, when necessary, relaying (adjusting the aim) of the guns. A salvo can be fired by entire batteries or by individual turrets or mounts. To conserve ammunition, reduce the intervals between salvos or speed up the firing of ranging shots, partial salvos or split salvos may be fired. (*US Navy*)

This photograph shows the turret officer's compartment of a 16-inch main gun battery on a preserved *Iowa*-class battleship. The compartment was located at the rear portion of the turret and was separated from the three individual gun pits that contained the breech end of the 16-inch guns by a flameproof bulkhead. This was done to protect the turret officer's compartment personnel if an accident or enemy fire destroyed or damaged one of the three gun pits in the turret. The large cylinder to the right of the picture is the 46-foot-long range-finder mounted in each 16-inch main gun battery turret on the US navy's fast battleships. (*Vladimir Yakubov*)

Seen here in its raised firing position with a dummy missile simulating the launching of a Tomahawk BGM-109 cruise missile is an armoured box launcher (ABL) on a preserved *Iowa*-class battleship. The ABL was fixed in position and was not aimed either right or left. The eight ABLs fitted to all four of the modernized *Iowa*-class battleships in the 1980s were armoured, not so much to protect them from enemy fire but from the blast and concussion generated when the battleships' main gun battery turrets were fired. (*Vladimir Yakubov*)

Leading a screening line in a carrier battle group in 1989 is the USS *New Jersey* with the USS *Missouri* behind it. Following the *Missouri* is the nuclear-powered, guided missile armed cruiser USS *Long Beach* (CGN-9). An external identification feature on the *New Jersey* that distinguishes it from its three 1980s-modernized sister ships is the shape of the electronic countermeasure system compartment located directly below the Mk 38 director. The configuration of the electronic countermeasure compartment on other battleships can be seen on the following USS *Missouri*. (*US Navy*)

Berthed at Pearl Harbor since 1999 is the preserved USS *Missouri*, one of the top tourist attractions on the island of Oahu. Located directly behind the battleship's superstructure is the ship's foremast, topped with a horizontally-oriented black air-search radar antenna. Above the antenna is a very tall, black, vertically-oriented radio transmitter. The foremast straddles the forward stack and is also supported by the aft portion of the superstructure tower. (*US Navy*)

A Tomahawk BGM-109 cruise missile is shown here during the 1980s being launched from the USS *New Jersey*. Note the sailors clustered around the battleship's forecastle and fantail to view the launching of the cruise missile, a somewhat rare event in peacetime due to the cost of the weapon. The mounting of the cruise missile launchers on the *Iowa*-class battleships was one of the justifications for the reactivation of the ships by their supporters, as it was claimed that none of the US navy's other surface ships had the space to carry them in sufficient numbers. (*US Navy*)

Shown at sea sometime in the 1980s is the USS *Wisconsin*. The battleship lost a portion bow in a collision with a US navy destroyer in May 1955 during a training cruise. To speed up the ship's return to service, it was decided to take a 120-ton, 68-foot section of the bow from the never-completed *Iowa*-class battleship USS *Kentucky* (BB-66) and craft it on to the front of the *Wisconsin* in place of the damaged bow. As with its three *Iowa*-class sisters, the battleship has been preserved as a museum ship and has been berthed at Norfolk, Virginia since 2001. (*US Navy*)

An impressive image of the USS *Texas* firing a salvo from its main gun battery turrets. The combined weight of the projectiles fired in a salvo from the battleship was 15,000lb. The battleship was fitted with various models of the 14-inch main battery guns during its time in service. The tripod masts and the missing flag (admiral's) bridge mark the photograph as having been taken between 1926 and 1927. (*National Archives*)

The USS *Texas* is shown here in this photograph dated 12 February 1930. One of the modernization improvements made to the ship, and to other dreadnought-type US navy battleships in-between the First and Second World Wars, was the addition of new amidships armoured casemates as seen in this port-side photograph. These armoured casemates housed six 5-inch secondary guns (three on either side of the ship). (*National Archives*)

The crew of a 5-inch secondary gun on an unnamed US navy battleship is shown preparing to load a projectile into the breech end of the weapon. Another member of the gun crew is preparing to open the weapon's breech mechanism, the mechanical device at the rear end of every breech-loading gun. Due to the weight of the complete round, the 5-inch ammunition was separate-loading with the projectile going in first, followed by a silk powder bag. (*National Archives*)

To sharpen the ammunition-handling skills of the crews on the 5-inch secondary guns mounted on US navy battleships, loading practice machines were mounted on the ships as seen in this photograph from the USS *Texas*. The weapon's projectiles weighed up to 55lb and the silk powder bag 24.5lb. Dummy projectiles and powder bags are seen on the main deck. (*Texas SHS*)

On the upper deck (a partial deck above the main deck amidships) of the USS *Texas* can be seen a 5-inch secondary gun in the foreground. In US navy manuals the weapon was designated the 5"/51. The suffix '51' stood for the gun's calibre (barrel length), which was derived from multiplying the bore size by the calibre (5 inches × 51 = 255 inches or 21 feet 3 inches). Behind the 5"/51 are two examples of the 3"/50 secondary gun. (*Texas SHS*)

On board an unnamed US navy battleship is a 5-inch secondary gun with a sailor taking aim through the weapon's optical sight. In the background a sailor stands ready to load a projectile into the gun and another prepares to swab (clean) the gun's barrel after firing. The manual aiming of a gun on any ship is extremely difficult as the vessels roll from side to side and pitch (the vertical rise and fall of a ship's bow and stern fore and aft). This tilts a weapon's trunnions (pivots on which the gun moves in elevation) out of level, causing errors in elevation and train (azimuth). (*National Archives*)

Being hoisted on board a US navy ship is a 5-inch secondary gun designated the 5"/25 in US navy manuals, which meant it had a barrel length of 125 inches or 10 feet 5 inches. Unlike the 5"/51 secondary gun that it replaced on US navy battleships, it was intended from the beginning as a dual-purpose weapon suitable for engaging both surface and aerial targets. Design work on the 5"/25 began in 1921 and it began appearing on US navy battleships in 1926. Because it was envisioned that it would take barrage fire to destroy incoming enemy aircraft, the 5"/25 fired fixed (one-piece) ammunition to speed up the weapon's rate of fire, rather than the more cumbersome separate-loading ammunition employed on the 5"/51. (*National Archives*)

(*Opposite*) Sometime during the Second World War the crews of the 3"/50 secondary guns aboard the USS *Texas* are shown engaging aerial targets. Note the ammunition storage lockers next to the guns. The 3"/50 secondary gun was adopted by the US navy during the First World War. The prototype weapon was installed on the USS *Texas* in 1916. By 1918 there was a four-gun battery of 3"/50 secondary guns on every US navy battleship, with this number being doubled in postwar refits. (*Texas SHS*)

(*Above*) A photographer has caught the firing of a 3″/50 secondary gun on the USS *Texas* during a training exercise. Unlike the 5″/51 secondary gun intended only as an anti-surface weapon employing separate-loading ammunition, the 3″/50 was a dual-purpose gun that could be employed against surface as well as aerial threats and employed fixed ammunition, resulting in a higher rate of fire. (*Texas SHS*)

AIRCRAFT CREW WATCHING FOR SUBMARINES

During the First World War the US navy mounted a single 3"/50 secondary gun on the top of each of the USS *Texas*'s two boat crane kingposts, as shown here, for a maximum field of fire. Note the anti-splinter matting on the exterior of the gun tub and the ammunition lockers attached to the gun tub. The 3"/50 secondary gun had a muzzle velocity of 2,700 fps and a rate of fire of between twelve to fifteen rounds per minute. (*Texas SHS*)

A picture taken either in 1943 or 1944 shows the USS *Texas* now fitted with a large number of newly-introduced small-calibre anti-aircraft guns. These include the Swiss-designed 20mm Oerlikon in single mounts and the Swedish-designed 40mm Bofors gun arranged in quadruple mounts. The addition of these weapons to the *Texas* and other dreadnought-type US navy battleships that saw service during the Second World War illustrated the serious threat posed by Japanese aerial attacks. (*National Archives*)

The 20mm Oerlikon anti-aircraft gun, seen here in this labelled illustration from a manual, was approved for service by the US navy in November 1940. An American-built version was first test-fired in June 1941 and went through a number of models. The most numerous variant in US navy service during the Second World War was designated the 20mm Single Mount Mk 4 and only fired in full automatic. The weapon's realistic effective range was approximately 1,000 yards.

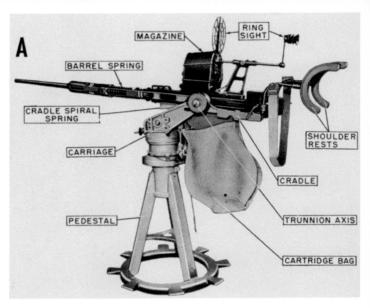

The 20mm Oerlikon anti-aircraft gun adopted by the US navy for its battleships replaced the .50 calibre water-cooled machine gun, seen here at an onshore training school surrounded by young Marines. The weapon's official designation was the .50 calibre Model 1921 and it was adopted by both the US navy and US army in 1929. By the early 1930s a US navy battleship was authorized a complement of eight .50 calibre Model 1921 machine guns. These weapons lasted in service on some battleships until 1942. (*National Archives*)

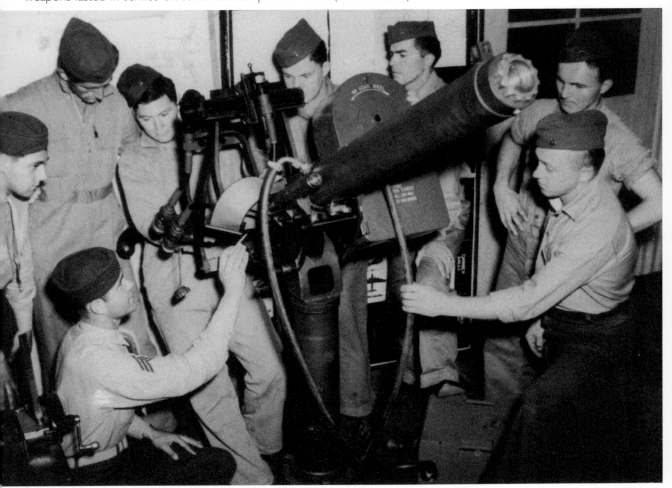

Another replacement for the .50 calibre Model 1921 machine guns mounted on board US navy battleships was the water-cooled quadruple 1.1-inch machine cannon seen here. The design of the weapon had begun in 1929 and testing of a single-barrel version started in 1931. The US navy approved the test results and ordered a power-operated quadruple version into production in 1934. However, by the time the weapon reached widespread service in 1940 it was obsolete against the then current generation of dive-bombers. (*National Archives*)

The replacement for the quadruple 1.1-inch machine cannon on US navy battleships starting in 1942 was the water-cooled quadruple 40mm Bofors gun, seen here in a labelled illustration from a manual. The weapon was officially designated the 40mm Quad Mount Mk 2. The Swedish-designed weapon first appeared on the USS *Wyoming* (BB-32) and was eventually fitted to every US navy battleship that saw service during the Second World War. The entire mount weighed approximately 13,000lb or about 6 tons.

FLASH HIDERS CARRIAGE GUNS WATER JACKETS ELEVATING SI
TRAINING SIGHT POINTER'S HANDWHE
LAG METER POINTER'S SE
TRAINER'S HANDWHEEL
TRAINER'S SEAT
FIRING STOP MECHANISM FIRING PED
POWER UNIT (TRAIN POWER DRIVE) ELEVATION POW DRI
CASE DISCHARGE CHUTES FIRING MOTOR STARTE

Chapter Three

Super Dreadnoughts

Authorized by Congress in 1911 were the two battleships of the *Nevada* class, which comprised the USS *Nevada* (BB-36) and USS *Oklahoma* (BB-37). Both ships were commissioned in 1916 and had a full load displacement of 28,400 tons, making them just slightly heavier than the *New York*-class battleships that went before them. They were 583 feet long and had a beam of 95 feet 3 inches. Crew complement on the ships was 864 officers and enlisted men in peacetime and 1,598 in wartime.

Despite outward similarities to their direct predecessors, the *Nevada*-class battleships were much improved over the dreadnoughts that came before them; so much so that naval historians often refer to them and the five follow-on classes of battleships as either 'super dreadnoughts' or 'second-generation dreadnoughts'.

The US navy also referred to the two *Nevada*-class battleships and the ten battleships that followed them into service as 'standard-type battleships'. This meant that all of them had the same general operational characteristics such as handling and maximum speed, allowing them to serve together in battle when called upon.

The *Nevada*-class ships retained the ten 14-inch main battery guns, as had the previous *New York* class. The difference was that the ten 14-inch main battery guns were not divided between only four turrets, two forward of the superstructure and two aft. The main gun battery turrets located on the weather/protective deck forward and aft of the ship's superstructure were armed with three 14-inch guns each, while the super-firing turrets behind and above them were armed with only two 14-inch guns each. The *Nevada*-class battleships were the first to have main gun battery turrets fitted with three guns.

The ships' secondary battery originally consisted of twenty-one casemated 5-inch guns, later reduced to twelve as their location in the hull meant they were wet a great deal of the time and therefore unusable. Eventually the ships had their casemated 5-inch guns replaced by two different generations of dual-purpose 5-inch guns; in their final version, protected by armoured mounts.

With the *Nevada* class the US navy adopted a new concept of armouring its battleships, referred to as the 'all-or-nothing' approach that was continued on all subsequent battleship classes. It meant that the ship's armour protection was now only concentrated around the most important and vulnerable aspects of its design, such as the command and control facilities, the propulsion plant and the ammunition and propellant storage areas. These areas of the ship were concentrated around the centre section of the vessel and became known as the 'armoured citadel'. Areas outside the citadel, such as the crew's berthing areas, were left unprotected. This new all-or-nothing approach was in contrast to the previous approach on early US navy battleships and dreadnoughts that had varying thicknesses of armour arrayed around the entire ship.

The sloping vertical armoured faces of the main gun battery turrets on the *Nevada*-class battleships were 18 inches thick. Hull belt armour had a maximum thickness of 13.5 inches, with the protective decks being 3 inches thick. The conning towers were built of armour 11.5 inches thick. The ships' lower hulls were double-bottomed and protected by external torpedo bulges.

As the ranges at which battleships could now engage each other grew, the danger from long-range plunging fire increased. As a result, the *Nevada*-class battleships were the first to have a second armoured deck placed below the uppermost protective deck, referred to as a 'splinter deck'. This would be a feature seen on all subsequent US navy super dreadnought classes with the two armoured decks ranging in maximum thickness from 4.5 to 5 inches.

Another marker for the super dreadnoughts in US navy service was having only fuel oil-fired boilers driving the ships' engines. Previous battleships, such as one of the two ships of the *Delaware*-class battleships and the follow-on *Florida* and *Wyoming* class had both coal and fuel oil-fired boilers. This mixture of boiler types was driven by design problems with the early fuel oil-fired boilers that were eventually resolved by the *Nevada*-class battleships.

A big advantage of switching from coal-fired boilers to all fuel oil-fired boilers on US navy super dreadnoughts was a reduction in the number of men needed to service the ships' engines. Another advantage of fuel oil-fired boilers is that they were more efficient than the coal-fired boilers, which meant that fewer of them were needed on battleships. Fuel oil was also cleaner and easier to handle than coal on board ships. By switching to fuel oil-fired boilers for its ships, the US navy could do away with its worldwide coaling stations. This was a trend then being followed by all the major navies of the world.

The USS *Nevada* was fitted with steam turbine engines, while the USS *Oklahoma* had triple-expansion reciprocating engines, the last US navy battleship to be fitted with these. Both ships had a maximum speed of 21 knots. The steam turbine engines on the USS *Nevada* were the first geared steam turbine engines fitted to a US navy battleship, which then became a standard feature on many of the follow-on classes. The Royal Navy switched to geared steam turbine engines for all their new ships in 1915.

Both *Nevada*-class battleships saw service during the First and Second World Wars with the *Nevada* heavily damaged, with 50 men killed and 109 wounded, during the Japanese surprise attack on the US navy base at Pearl Harbor on 7 December 1941. The battleship was later refloated and returned to service by May 1943.

The USS *Oklahoma* was not so fortunate during the Pearl Harbor attack as the USS *Nevada* and capsized after being struck by Japanese aerial torpedoes and bombs with a loss of 415 men killed. The ship was eventually refloated but a decision was made not to rebuild it and the vessel had its final decommissioning in 1944. On the way to the scrapping yard the ship sank while under tow.

The *Nevada* saw combat throughout the war in the Pacific from May 1943 until the Japanese surrender in August 1945. It was one of four US navy battleships employed as target ships during Operation CROSSROADS in July 1946. It survived the detonation of two atomic bombs, only to be later sunk by the US navy off Hawaii in July 1948 during weapon-testing trials.

Pennsylvania Class

Following on the heels of the *Nevada*-class battleships came two *Pennsylvania*-class ships: the USS *Pennsylvania* (BB-38) and the USS *Arizona* (BB-39). They were basically slightly enlarged *Nevada*-class variants authorized in 1912 and 1913 respectively and commissioned in 1916. They were armed with twelve 14-inch guns divided between four centreline main gun battery turrets, with each turret housing three guns and referred to as 'triple turrets'.

The secondary battery on the *Pennsylvania*-class battleships originally consisted of twenty-two 5-inch guns intended as torpedo-boat protection, as with many of the super dreadnoughts and dreadnoughts that went before them. However, during their modernization process in the 1920s, the ships were fitted with 5-inch guns that could engage both surface and aerial targets. These dual-purpose 5-inch guns were described in US navy documents as the Mark 5"/25. The latter number is the calibre, which when multiplied by the diameter of the bore provides the barrel length of the weapon; in this case 125 inches (10 feet 5 inches).

The 5"/25 dual-purpose guns were intended as replacements for the pedestal-mounted 3-inch dual-purpose guns, referred to as the 3"/50, that had first appeared on US navy ships in 1917 during the First World War. Barrel length on the weapon was 150 inches (12 feet 6 inches). Some of these 3"/50 dual-purpose guns lasted in service on US navy ships throughout the Second World War.

The *Pennsylvania*-class battleships had the same armour arrangement as seen on the previous *Nevada* class, except that the armour on the conning towers of the *Pennsylvania*-class ships was slightly thicker at 16 inches. As with the preceding battle-ships, the lower hull was protected by a double bottom and external torpedo bulges.

Initial full load displacement of the *Pennsylvania*-class battleships was 32,657 tons, with the USS *Pennsylvania* growing to 40,330 tons following modernization in the 1920s. The USS *Arizona* went up to 37,654 tons upon modernization from 1929 to 1931. Both ships had a length of 608 feet and a modernized beam of 100 feet on the *Pennsylvania* and 106 feet 2 inches on the *Arizona*. The *Pennsylvania*-class battleships had a crew complement of 915 officers and enlisted men and 1,574 during wartime. They both had geared steam turbine engines powered by fuel oil-fired boilers, giving them a top speed of 21 knots.

A key external spotting feature of the modernization process applied to many of the US navy's super dreadnoughts and the previous dreadnoughts was the replacement of the original tall cage masts with heavy-duty tripod masts. The main battery guns on the super dreadnoughts and dreadnoughts were also provided with more elevation to increase the range of the weapons. All of the secondary battery guns were now located on or above the hulls of the super dreadnoughts and dreadnoughts.

The USS *Pennsylvania* and USS *Arizona* both served during the First World War and found themselves at Pearl Harbor on 7 December 1941 when the Japanese attacked. The USS *Pennsylvania* was in dry dock at the time and was protected from aerial torpedoes but lost 15 men killed, 14 missing and 38 wounded to enemy bombs and strafing.

The USS *Pennsylvania* saw service throughout the remainder of the Second World War. It was seriously damaged off the island of Okinawa on 12 August 1945 by a Japanese torpedo and then out of service for the remainder of the war. Postwar, it became a target ship during the atomic tests of Operation CROSSROADS. Badly damaged during testing, she was decommissioned in August 1946 at Kwajalein Lagoon and later scuttled at the same location in February 1948.

The USS *Arizona* sank on its keel at Pearl Harbor on 7 December 1941 after a number of enemy bomb hits, one of which penetrated its forward powder (propellant) magazine and exploded, killing the ship's captain and the rear admiral aboard the ship. Another 1,104 officers and enlisted men also perished in the same explosion. The ship was never raised and became a memorial for all the American military personnel killed during the Japanese attack.

New Mexico Class

About the same size and full load displacement of the two *Pennsylvania*-class battleships were the three *New Mexico*-class battleships that followed them into service: the USS *New Mexico* (BB-40), the USS *Mississippi* (BB-41) and the USS *Idaho* (BB-42). Authorized by Congress in 1914, the year that the First World War began, the three ships were commissioned between 1917 and 1918 and had a full load displacement of 33,000 tons. Length for the ships was 624 feet, with a beam of 97 feet 4 inches.

The main gun batteries on the *New Mexico*-class battleships consisted of twelve 14-inch guns divided between four centreline turrets, two forward of the ships' superstructures and two aft, each armed with three guns. These 14-inch main battery guns were 58 feet long, or 50 calibre. Those on the prior *Pennsylvania*-class battleships were only 52 feet 5 inches long, or 45 calibre. A longer gun tube means an increase in range and typically, a greater ability to penetrate armour.

The *New Mexico*-class battleships' secondary battery armament originally consisted of 22 5-inch casemated guns, later dropped to 14 and then 12. By the time the casemated 5-inch guns had been reduced to 12, the ship's secondary battery armament was upgraded with 8 dual-purpose 5-inch guns. These would be supplemented during the Second World War by a large number of small-calibre anti-aircraft guns.

Armour protection on the *New Mexico*-class battleships was the same as on the previous *Pennsylvania* class except for the conning tower that had a maximum armour thickness of 11.5 inches. The hulls of the *New Mexico*-class ships were protected by double bottoms and external torpedo bulges.

Of the three battleships in the *New Mexico* class, the USS *New Mexico* had turboelectric drive engines, powered by fuel oil-fired boilers. Its sister ships, the USS *Mississippi* and USS *Idaho*, retained geared steam turbine engines powered by fuel oil-fired boilers. The turboelectric drive engines on the USS *New Mexico* were eventually replaced with geared steam turbine engines in the 1930s, powered by fuel oil-fired boilers. The maximum speed on all three *New Mexico*-class battleships, no matter the type of engine, was 21 knots.

Crew complement for the *New Mexico* class in peacetime ranged from 1,081 to 1,804 officers and enlisted men, and in wartime from 1,560 to 1,600. The three *New Mexico*-class battleships did not see any action during the First World War but saw productive employment during the Second.

The three *New Mexico*-class battleships all saw their final decommissioning in 1946. The USS *New Mexico* was sold for scrapping in 1947, as was the USS *Idaho*. The USS *Mississippi* was decommissioned as a battleship on 2 February 1946. However, it would see a second life as a test ship for new weapon systems under the designation EAG-128 and later just AG-128. The letter prefix 'AG' stands for auxiliary general. As the AG-128, the ship had its final decommissioning in September 1956.

Tennessee Class

Also authorized by Congress in 1914 were the two super dreadnought battleships of the *Tennessee* class, comprising the USS *Tennessee* (BB-43) and the USS *California* (BB-44). They were commissioned in 1919, too late to see action during the First World War. Both were armed with the same 14-inch guns mounted on the previous

New Mexico class, divided between four centreline turrets each armed with three guns, two forward of the ships' superstructures and two aft.

Secondary battery armament on the *Tennessee*-class battleships consisted originally of 5-inch and 3-inch dual-purpose guns, the 3-inch guns eventually being replaced by additional 5-inch guns. Based on favourable Royal Navy experience gained during the Battle of Jutland, which took place from 31 May to 1 June 1916, the US navy adopted a centralized fire-control system to improve their efficiency for the secondary battery guns on the *Tennessee*-class battleships and the battleship classes that followed. With the outbreak of the Second World War the ships received an extensive array of small-calibre anti-aircraft guns.

The thickness of the armour on the *Tennessee*-class battleships was the same as that on the previous *New Mexico* class. Gone, however, were the external torpedo bulges on the lower hull. They were replaced on the *Tennessee*-class ships, and all the subsequent classes of US navy battleships, with internal torpedo belts, separated by armoured bulkheads consisting of layers of watertight compartments along the double bottoms of their hulls. Like the external torpedo bulges, the new internal torpedo belts were intended to absorb and dissipate the detonation of an under-water mine or a torpedo warhead.

Both the *Tennessee*-class battleships found themselves at the Pearl Harbor naval base when the Japanese military struck on 7 December 1941. As the USS *Tennessee* was inboard of the USS *West Virginia*, it was struck by only two armour-piercing bombs. It set sail back to the West Coast of the United States for rebuilding above the main deck two weeks after the Japanese attack, rejoining the US navy Pacific Fleet in August 1943 (it operated from February to August 1942 off the West Coast, prior to a refit). The USS *Tennessee* had an original full load displacement of 33,190 tons that later increased to 40,400 tons under modernization.

The USS *California* was not as lucky as its sister ship at Pearl Harbor and was sunk at its mooring by Japanese torpedoes and a bomb, with the loss of ninety-four killed and fifty-one wounded. The ship should have survived the attack but all the interior watertight doors were open for a Sunday morning inspection. The ship was refloated in early 1942 and was rebuilt above the main deck at Bremerton, Washington. This took until January 1944 when it then rejoined the Pacific Fleet. The initial full load displacement of the *California* was 33,190 tons and 40,400 tons upon completing modernization.

The most dramatic external feature that highlighted the wartime rebuilding of the two *Tennessee*-class battleships, besides the heavy emphasis on secondary battery anti-aircraft guns, was the removal of the two tall cage masts. In their place was a massive superstructure bridge assembly near the front of the ship, just behind the forward main battery gun turrets. The superstructure provided space for the ship's 'nerve centre', for within its collection of levels (floors) and compartments were the personnel and instruments required for its command, communication with other parts of the ship as well as with other ships, and for the control of gunfire.

Peacetime crew complement on the two *Tennessee*-class battleships was 1,083 officers and enlisted men, with their wartime complement rising to 1,407. The USS *Tennessee* and USS *California* had their final decommissioning in 1947 and were both sold for scrapping. Length of the ships was 624 feet 6 inches, with a modernized beam of 114 feet. Like the USS *New Mexico* (BB-40), the two *Tennessee*-class ships had turboelectric drive engines powered by fuel oil-fired boilers and they would retain these engines throughout their service lives. The maximum speed of the *Tennessee* class was 21 knots.

Colorado Class

Authorized in 1916 by Congress were the four battleships of the *Colorado* class: the USS *Colorado* (BB-45), the USS *Maryland* (BB-46), the USS *Washington* (BB-47) and the USS *West Virginia* (BB-48). The USS *Washington* was cancelled in 1922, with the other three ships being commissioned between 1921 and 1923.

The biggest change between the *Colorado*-class battleships and those that went before them was the replacement of the 14-inch main battery guns with 16-inch main battery guns. Instead of twelve 14-inch main battery guns divided between four centreline turrets, each armed with three guns, the *Colorado*-class battleships' four centreline turrets housed two 16-inch guns each. They would be the last US navy battleship class with four main gun battery turrets.

Secondary battery armament on the *Colorado*-class battleships consisted originally of 5-inch and 3-inch dual-purpose guns, the 3-inch versions eventually being replaced by additional 5-inch guns. With the outbreak of the Second World War the ships received an extensive array of small-calibre anti-aircraft guns.

The armour arrangement of the *Colorado*-class battleships was the same as on the earlier *Tennessee* class, including a double bottom and a torpedo belt.

The *Colorado*-class ships had an original full load displacement of 33,590 tons that rose to 39,400 tons upon modernization. Ship length was 624 feet with a modernized beam of 108 feet. The crew complement of the ships ranged from 1,070 officers and enlisted men to 1,084. All three of the commissioned *Colorado*-class battleships had turboelectric drive engines powered by fuel oil-fired boilers, which they would retain until being scrapped. The maximum *Colorado*-class battleship speed was 21 knots.

Of the three *Colorado*-class battleships commissioned, two were at Pearl Harbor on 7 December 1941: the USS *Maryland* and the USS *West Virginia*, with the former suffering only very minor damage. The *West Virginia* was sunk at its mooring by a combination of enemy bombs and aerial torpedo strikes. Refloated in 1942, the ship was rebuilt above the main deck at Puget Sound Naval Shipyard located in the state of Washington and returned to fleet service in 1944. All three *Colorado*-class battleships were decommissioned in 1947 and sold for scrapping in 1959.

Five of the US navy's super dreadnought battleships damaged or sunk at Pearl Harbor by the Japanese surprise attack and placed back into service had their revenge on the night of 25 October 1944, during the Battle of Surigao Strait. In that engagement the USS *Pennsylvania* (BB-38), USS *Tennessee* (BB-43), USS *California* (BB-44), USS *Maryland* (BB-46) and USS *West Virginia* (BB-48) waited in ambush for a Japanese naval force that included two battleships: IJN *Yamashiro* and IJN *Fusō*. In the ensuing battle, the Japanese navy lost both its battleships (already damaged by torpedo strikes from US navy destroyers supporting the battleships) with no losses to the five Pearl Harbor battleship survivors. They were aided in that battle by the USS *Mississippi* (BB-41) that had not been at Pearl Harbor on 7 December 1941.

South Dakota Class

Reflecting the continued provocations by Imperial Germany against a then-neutral United States, American President Woodrow Wilson, who had originally preferred not to become involved in the First World War, endorsed the Naval Act of 1916, also referred to as the 'navy second-to-none act'. It called for a greatly expanding building programme that resulted in ten new battleships built for the US navy in a three-year period. By late 1918 the US navy's leadership was calling for another twelve battleships to be built on top of the ten already planned.

In accordance with the Act, the American Congress authorized a number of battleships between 1916 and 1919. These included the six battleships of the *South Dakota* class: the USS *South Dakota* (BB-49), USS *Indiana* (BB-50), USS *Montana* (BB-51), USS *North Carolina* (BB-52), USS *Iowa* (BB-53) and USS *Massachusetts* (BB-54). These ships were to have as their main gun battery twelve 16-inch guns divided between four centreline turrets, two forward of the ship's superstructure and two aft, each armed with three guns. Secondary battery armament was to be a combination of 5-inch and 3-inch dual-purpose guns.

The armour arrangement of the planned *South Dakota*-class battleships was to be a bit thicker in some locations than that of the previous classes of super dreadnought battleships, with two armoured hull decks below the uppermost protective hull decks. With a three armoured deck arrangement, the uppermost armoured deck was referred to as the 'bomb deck' with the armoured deck below it designated the 'protective deck' and the lowest armoured deck becoming the 'splinter deck'. This arrangement of armoured decks would become a standard design feature on follow-on classes of US navy battleships.

The *South Dakota*-class battleships were to have a full load displacement up to 43,200 tons, with a crew complement of 1,191 officers and men. Length was to be 684 feet, with a beam of 106 feet. All were to have turboelectric engines powered by fuel oil-fired boilers. The maximum speed of the *South Dakota*-class battleships was intended to be 23 knots.

The keel for five of the *South Dakota*-class ships was laid down in 1920 with the keel for the USS *Massachusetts* laid down the following year. None of the *South Dakota*-class vessels were ever commissioned due to the Washington Naval Treaty of 1922 and subsequent treaties agreed by the victors of the First World War to curtail a naval arms race by limiting the number of battleships by tonnage limitations. These included the United States, Great Britain, Japan, Italy and France.

Treaty-Imposed Limitations

To accomplish the goals of the Washington Naval Treaty, older generation battleship classes of the various navies were decommissioned and scrapped and new battle-ship classes cancelled for ten years, later increased to a fifteen-year moratorium on new battleship construction based on follow-on naval arms control treaties. When new battleships were eventually built, they were limited to a standard load displacement of 35,000 tons. They were to be armed with main gun batteries no larger than 16-inch.

All the various major navies – American, British and Japanese – were loath to see the various arms control treaties put into effect. However, their governments knew that a continued arms race of battleships would be ruinous to their respective economies. In both the United States and Great Britain, public support of their respective navies continuing to build additional battleships had quickly evaporated following the end of the First World War.

To meet the mandates of the Washington Naval Treaty and follow-on treaties, the US navy had the *South Dakota*-class super dreadnought battleships cancelled before their completion. Their 16-inch guns went to the US army coast artillery branch for coastal defence on both the east and west coasts of the United States. The US navy also had twelve of its pre-dreadnought battleships and three of its dreadnought battleships decommissioned and scrapped. Provisions were made in the various treaties to modernize existing battleships with added protection from aircraft and submarines.

Shown in its original configuration is the USS *Nevada* (BB-36) with an open navigation bridge and two cage masts. The ship was constructed at the Fore River Company, Quincy, Massachusetts. Reflecting its status as the first so-called 'super dreadnought'-type battleship in US navy service, it boasted fuel oil-fired boilers when built. Another unseen interior feature on the super dreadnought-type battleships was the armouring of only the most critical areas of the battleship, such as the engine rooms and the ammunition magazines, rather than the entire ship. This protected space on the ships was referred to as the 'armoured citadel'. (*National Archives*)

(*Opposite*) This aerial photograph shows the modernized super dreadnought USS *Nevada*. It is now fitted with tripod masts instead of the original cage masts and features an enclosed navigation bridge. The key external identifying feature of the super dreadnought-type battleship in US navy service was two main gun battery turrets (Nos 1 and 4) armed with three 14-inch guns. The super-firing main gun turrets (Nos 2 and 3) retain the two 14-inch gun arrangement seen on all the dreadnought-type battleships that came before it. (*National Archives*)

(*Above*) Berthed at the US navy base Pearl Harbor, Hawaii on the morning of 7 December 1941 when Japanese aircraft struck, the USS *Nevada* was the only battleship that managed to make it under way. However, once under way, it became the centre of attention for the Japanese aerial attackers who concentrated their attention on the ship in the hope of sinking it in the main entrance to the American naval base. Having been struck by one aerial torpedo and up to ten bombs, the battleship was ordered beached, as seen in this photograph. A harbour tug is moored to her port bow helping to fight fires. (*National Archives*)

Pictured is the USS *Nevada* in 1945 before the end of the Second World War. Refloated at Pearl Harbor in February 1942, the battleship made its way to Puget Sound Navy Yard where it went through a major overhaul and modernization, completed in October 1942. Gone were the two tripod masts, replaced by a massive centreline tower superstructure. It has been fitted with numerous radar antennas mounted on its superstructure. (*National Archives*)

Another key external identifying feature of the major overhaul and modernization process done on the USS *Nevada* after Pearl Harbor was the removal of the ship's 5"/51 and 5"/25 secondary battery guns. In their place the battleship received eight powered armoured mounts, two of which are seen amidships on the port side of the ship's superstructure in this 1944 photograph. Each of these armoured turrets was armed with two dual-purpose 5-inch guns. The two guns and armoured mount were designated the 5"/38 Twin Mount 32. (*National Archives*)

Pictured is a close-up example of a 5"/38 Twin Mount 32 on a US navy battleship. The twin guns fired semi-fixed ammunition at a rate of fire of up to twenty-two rounds per minute per gun for a short period of time. The typical rate of fire for the two guns on the 5"/38 Twin Mount 32 was approximately fifteen rounds per minute per gun. The maximum horizontal range of the guns was a little over 10 miles. When the guns were at their maximum elevation they could engage aircraft up to an altitude of 37,000 feet. (*US Navy*)

From a US navy manual appears this labelled overhead illustration of a 5"/38 Twin Mount 32. Ammunition and powder cartridges for the gun room (also known as the gun house) came from magazines located below it by way of power-driven ammunition hoists. The ammunition was semi-fixed, consisting of a 53- to 55-lb projectile and a 28-lb cartridge case containing the propellant. Ammunition types included high-explosive fragmentation rounds for aerial targets.

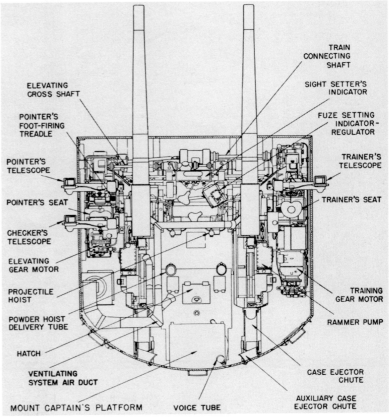

ELEVATING CROSS SHAFT

POINTER'S FOOT-FIRING TREADLE

POINTER'S TELESCOPE

POINTER'S SEAT

CHECKER'S TELESCOPE

ELEVATING GEAR MOTOR

PROJECTILE HOIST

POWDER HOIST DELIVERY TUBE

HATCH

VENTILATING SYSTEM AIR DUCT

MOUNT CAPTAIN'S PLATFORM

VOICE TUBE

TRAIN CONNECTING SHAFT

SIGHT SETTER'S INDICATOR

FUZE SETTING INDICATOR-REGULATOR

TRAINER'S TELESCOPE

TRAINER'S SEAT

TRAINING GEAR MOTOR

RAMMER PUMP

CASE EJECTOR CHUTE

AUXILIARY CASE EJECTOR CHUTE

A picture taken inside the armoured turret of a 5"/38 Twin Mount 32 on a US navy battleship shows three members of the gun crew. The crew typically included two projectile-men and two powder-men. They stood by their respective guns and transferred the projectiles and the cartridge cases in a single movement from ammunition hoists to the rammer tray of the guns. The semi-fixed rounds were then loaded into the chamber of the guns by power rammers. (US Navy)

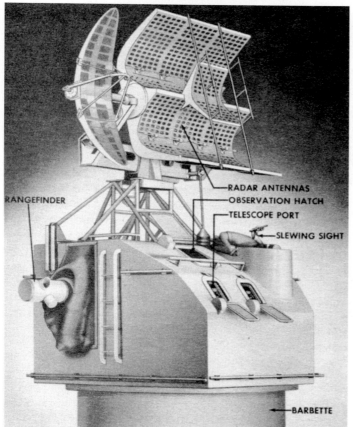

RANGEFINDER

RADAR ANTENNAS
OBSERVATION HATCH
TELESCOPE PORT
SLEWING SIGHT

BARBETTE

As some of the US navy's dreadnought- and super dreadnought-type battleships went through a post-Pearl Harbor major overhaul and modernization, their fire-control systems were updated with up to four of the armoured Mk 37 directors shown here. These 360° traversable directors were mounted on cylindrical barbettes as high as possible on a battleship's superstructure. Projecting out of either side of the director was the stereoscopic range-finder and on its roof were the radar antennas. The Mk 37 director provided automatic control for battleship secondary batteries including the 5"/38 Twin Mount 32 and could do the same for the 40mm anti-aircraft guns, if needed. (US Navy)

A young US navy lieutenant looks out of the overhead hatch on his armoured Mk 37 director. The gunfire director was first envisioned in 1936 and successfully tested in 1939. It was the first gunfire director for the US navy designed from the ground up to be equipped with radar. With the advent of the Mk 37 director the mechanical computer system that had been mounted inside earlier directors was moved into the armoured citadel of the ships. (*National Archives*)

All of the information provided by the Mk 37 director mounted on US navy battleships was transmitted to an electro-mechanical device referred to as the 'computer' or 'range-keeper'. It processed the information to determine the correct sight angle and deflection for the battleship's secondary gun batteries so they could accurately fire on the chosen target. Before the advent of the electro-mechanical computer on these battleships, manually-operated range tables were employed along with plotting tables. Pictured are three US navy sailors manning a Mk I Fire Control Computer. (*National Archives*)

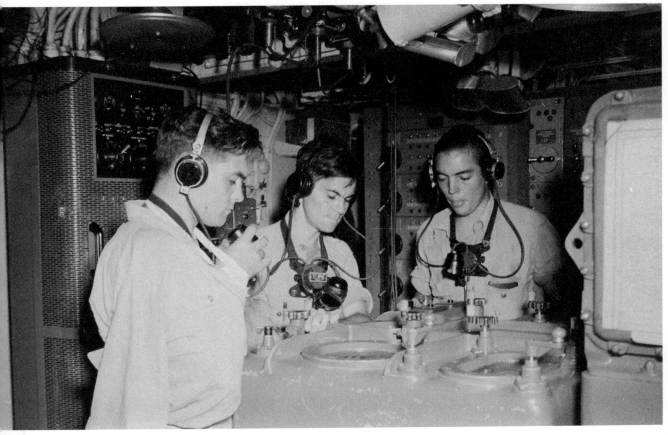

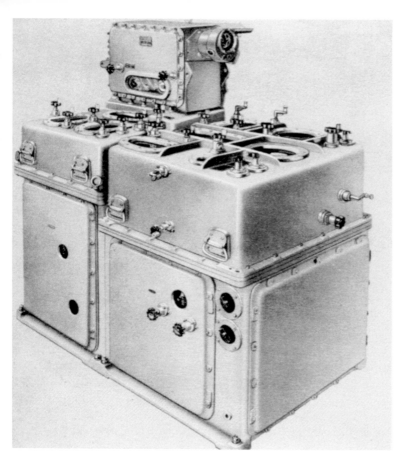

From a US navy manual comes this image of a fire-control computer employed by US navy battleships during the Second World War. It obtained all the required information from the ship's sensors and manually inputted information from its operators, such as wind and air density, before deciding on a firing solution for the secondary gun batteries. All the Mk 37 director personnel had to do was let the computer automatically operate their controls while they watched through their eyepieces, ready to apply corrections if the target or targets changed course or speed or if the computer was making an obvious error.

As the threat from aerial attack grew during the First World War, the US navy responded by adding secondary dual-purpose battery guns to the USS *Oklahoma*, as it would with other battleships employed during the conflict. Other wartime improvements to the ship included reducing the number of hull-mounted casemated 5"/51 guns and moving the remaining guns amidships into armoured casemates on either side of the superstructure. (*National Archives*)

Shown at the Philadelphia Navy Yard in 1929 is the USS *Oklahoma* (BB-37) during the final stages of modernization when its cage masts were replaced by tripod masts. The battleship was built by the New York Shipbuilding Corporation, Camden, New Jersey and was the sister ship of the USS *Nevada* (BB-36). During the First World War the *Oklahoma* served alongside the Royal Navy's Grand Fleet but saw no action other than escorting an allied convoy.

(*National Archives*)

When the Japanese attack struck the naval base at Pearl Harbor, Hawaii, on 7 December 1941 the *Oklahoma* was moored outboard alongside the USS *Maryland* (BB-46). It was quickly struck by at least five aerial torpedoes that led to the battleship capsizing, as seen in this picture, with only the ship's starboard side and part of the keel exposed above the water and 429 crew members perishing in the attack. Visible behind the capsized *Oklahoma* is the *Maryland*. (*National Archives*)

(*Opposite*) During a training exercise conducted during the 1930s, the main gun battery of the USS *Oklahoma* is pointed starboard in this picture. During the ship's modernization in 1927 its secondary battery was improved by the addition of eight 5"/25 dual-purpose guns. The main gun battery turrets, as with other modernized US navy battleships, were modified to increase their maximum elevation from 15 to 30 degrees in order to increase their range. In addition, anti-torpedo bulges were added to the hull, which increased the ship's beam from 95 feet 3 inches to 108 feet. (*National Archives*)

Plans for salvaging the capsized USS *Oklahoma* began in July 1942 and involved erecting twenty-one large winches onshore that were intended to right the battleship so it could be refloated. Prior to the righting of the ship, it was lightened by removing some of its fuel oil, ammunition and machinery. The job of refloating the ship was completed in November 1943 and it was moved to a dry dock at Pearl Harbor the following month, as seen in this picture of the now barnacle-encrusted hull. (*National Archives*)

Once in dry dock the refloated USS *Oklahoma* was made watertight by patching the large holes in its port side hull caused by the Japanese aerial torpedo hits. All of the ship's remaining fuel oil, ammunition and machinery were removed. The battleship was then towed to a pier at Pearl Harbor and moored. When it was decided that the elderly ship would not be rebuilt, it was decommissioned in September 1944 and all remaining armament and superstructure were removed, as is seen taking place in this picture. (*National Archives*)

The biggest difference between the two *Nevada*-class battleships and the two follow-on *Pennsylvania*-class battleships and all the super dreadnought-type battleships to follow was the addition of two more 14-inch guns, one each mounted in the Nos 2 and 3 main gun battery turrets of the ships. This can be seen on the No. 2 super-firing main gun battery turret on the forecastle of the USS *Pennsylvania* (BB-38) shown here, following its 1929 modernization. The battleship was built by the Newport News Shipbuilding Company, Newport News, Virginia. (*National Archives*)

A young sailor is shown taking the time to write a letter to his mother, or maybe a sweetheart, aboard the USS *Pennsylvania* in 1918. His location is in front of one of the ship's 5"/51 hull-mounted casemated guns. As originally built, the battleship was armed with twenty-two of the 5"/51 guns. Two were located in the forecastle deckhouse, two on either side of the hull near the stern of the ship and eight on either side of the weather deck below the forecastle deck. The weapon had a muzzle velocity of 3,150 fps. (*National Archives*)

Pictured firing a salvo from its main gun battery turrets is the USS *Pennsylvania* prior to its 1929–30 modernization, when it still had its cage masts. Because the battleship had been built with oil-fired boilers and no additional tankers could be spared, it was not assigned to serve with the Royal Navy's Grand Fleet during the First World War as were so many of the US navy's earlier dreadnought-type battleships. Instead the ship spent its time in training activities along the eastern coast of the US. (*National Archives*)

From 1922 until 1929 the USS *Pennsylvania* formed part of the US navy's Pacific Fleet until it departed for modernization at the Philadelphia Navy Yard in June 1929. During modernization its secondary battery of 5″/51 guns was eventually reduced from twenty-two to only twelve guns in 1929, six on either side of the superstructure amidships, in armoured casemates. In addition, its 3″/50 dual-purpose guns were replaced by eight 5″/25 dual-purpose guns. Following its modernization, as seen here, it returned to its Pacific Fleet duties. (*National Archives*)

In dry dock at Pearl Harbor on 7 December 1941 the USS *Pennsylvania* was protected from Japanese aerial torpedo hits but was struck by a number of enemy bombs. Total casualties during the attack were fifteen killed, fourteen missing in action and thirty-eight wounded. In October 1942 the ship went through a major overhaul and modernization process, during which it lost its tripod main mast but kept its tripod foremast, as seen in this 1944 photograph. It also acquired a redesigned superstructure mounting new fire-control equipment. (*National Archives*)

In August 1945 a Japanese plane managed to approach undetected the anchored USS *Pennsylvania* located off the island of Okinawa. It placed an aerial torpedo into the stern of the ship that created a 30-foot diameter hole in its lower hull. The torpedo hit killed twenty men and wounded another ten. It was only the hard work of the crew's damage control parties and the assistance of two tugboats that saved the battleship from sinking. This photograph shows the ship having the seawater that flooded its hull being pumped out. (*National Archives*)

(*Opposite*) After its modernization the USS *Arizona* returned to the West Coast of the United States in August 1931. The battleship's home port was San Pedro, California, the port district of Los Angeles. Visible in this stern view of the ship is a launching catapult on the fantail with an observation floatplane mounted on it. There is also a catapult on the battleship's No. 3 main gun battery superimposed turret with another observation floatplane mounted on it. (*National Archives*)

(*Above*) The USS *Arizona* (BB-39) seen here in this 1930s photograph was built at the New York Navy Yard, Brooklyn. The battleship did not serve overseas during the First World War and spent its time in training exercises along the eastern coast of the United States. From 1921 until 1929 the ship was based in Southern California. It was modernized between 1929 and 1931 and had its original cage masts replaced by tripod masts. (*National Archives*)

The USS *Arizona*, pictured here, remained at its home port of San Pedro until 1940. At the conclusion of its last training exercise, conducted off the Hawaiian Islands between April and May 1940, the battleship and others were ordered to remain at the naval base at Pearl Harbor. This was to deter the Japanese navy from any aggressive actions in the Pacific Ocean area, as the relationship between the United States and the Japanese governments had deteriorated. The contentious issues were primarily related to Japanese military activities in China. (*National Archives*)

(*Opposite*) Built by the New York Navy Yard, the USS *New Mexico* (BB-40) is shown here at its outfitting pier. It was the first commissioned of the three battleships in the *New Mexico* class and was a near repeat of the *Pennsylvania*-class battleships that entered service before it. Unlike all the US navy battleships that came before it that featured a ram bow, the *New Mexico* was fitted with a clipper bow which appeared on all subsequent US navy battleships. (*National Archives*)

(*Above*) Berthed at Pearl Harbor on the morning of 7 December 1941, the USS *Arizona* sustained a number of bomb hits from the attacking Japanese aircraft. The last bomb dropped penetrated the ammunition magazine located near turret No. 2 on the battleship's forecastle and in a massive explosion destroyed the forward portion of the ship's hull. This resulted in the conning tower and foremast collapsing forward, as seen in this photograph taken a few days after the attack. The casualty count for the battleship was 1,177 dead out of a crew of 1,512. (*National Archives*)

During the First World War the USS *New Mexico* served only along the eastern coast of the United States. After the conflict it sailed to France to return President Woodrow Wilson to America upon the conclusion of the Versailles Peace Conference. In the decades after the conflict the battleship served with both the Atlantic and Pacific fleets. It also visited various foreign ports to show the flag, a typical role performed by many US navy battleships prior to the First World War and in the decades before the start of the Second World War. (*National Archives*)

(*Opposite*) The USS *New Mexico* is seen here passing through the locks of the Panama Canal. The ship was 16 feet longer than the *Pennsylvania*-class battleships that preceded it and was the first in the US navy to feature a turbo-electric propulsion system. The ship visited Pearl Harbor in both 1925 and 1928 during training exercises. Such was the skill of the battleship's crew that it won the best in the fleet in gunnery, engineering and battle efficiency on a number of occasions between the 1920s and 1930s. (*National Archives*)

The USS *New Mexico* went through a modernization process between 1931 and 1933. It came out looking very different from the other super dreadnought and dreadnought battleships that underwent earlier modernization in the 1920s and 1930s. Rather than having its cage masts replaced by tripod masts which were in turn replaced by a tower superstructure in a second upgrade, the ship was fitted with a large superstructure tower as seen in this photograph dated 7 January 1938. The battleship did retain eight of its 5″/51 secondary battery guns in casemates amidships, four on either side of the superstructure. (*Maritime Quest*)

The USS *New Mexico* was based at the naval base at Pearl Harbor from December 1940 until May 1941 when she was once again assigned to the Atlantic Fleet. From 4 September 1940 until America officially entered the Second World War a few days after the Japanese attack on Pearl Harbor, the battleship was on neutrality patrol. In May 1942 the ship was provided with a large number of small-calibre anti-aircraft guns. The battleship was never fitted with the 5″/38 Twin Mount 32, as is evident from this photograph. (*Maritime Quest*)

Looking down from the superstructure tower of the USS *New Mexico* in June 1944 the weapons that made up the ship's small-calibre anti-aircraft battery are visible. To the lower left of the picture is the protruding barrel of the 20mm Single Mount Mk 4. Below that gun tub and to the right is a 40mm Twin Mount Mk 1. Below that and to its left is a 40mm Quadruple Mount Mk 2. By 1944, as the Japanese turned to aerial night attacks on US navy ships, the optically- and manually-aimed 20mm anti-aircraft guns fell out of favour. At that point in time, radar-guided 40mm anti-aircraft guns became more in demand. (*National Archives*)

Because the 40mm anti-aircraft guns mounted on US navy battleships during the Second World War were forced to engage their fast-moving aerial attackers at close ranges, the two most important pieces of information the gun crews required were the target's super-elevation and lead angle. To address this need the US navy developed a one-man-operated device, shown here, designated the Mk 51 director. It appeared in service in 1942 and was typically mounted above 40mm gun batteries for the best line of sight to the incoming target or targets and to allow the director operator to concentrate on aiming, away from the noise, smoke and chaos of the anti-aircraft batteries. (*Paul Hannah*)

The Japanese employment of suicide planes (Kamikaze), which began in earnest in October 1944, quickly demonstrated that the 20mm anti-aircraft gun lacked the knock-down power to deal with this new threat as it had only a very small high-explosive element. Pictured is the damage that a Japanese suicide plane caused in May 1945 to the starboard side of the USS *New Mexico* off the Ryukyu Islands when it struck a 20mm anti-aircraft gun position. *(National Archives)*

The USS *Mississippi* (BB-41) pictured here was constructed at the Newport News Shipbuilding Company, Virginia and was the second battleship in the *New Mexico* class of three ships. It was commissioned too late to see service during the First World War. It operated along the West Coast of the United States and was based at San Pedro, California from 1919 until 1922. Unlike its sister ship, the *New Mexico*, it was never fitted with turbo-electric engines; rather it had the more conventional steam turbine engines. *(National Archives)*

U.S.S. MISSISSIPPI
(BEFORE MODERNIZATION)

AV·19

Like its sister ship the USS *New Mexico*, the USS *Mississippi* went through a modernization process between 1931 and 1933 and the ship came out of the yard with a brand-new superstructure tower instead of the more conventional tripod masts, as shown in this picture. The battleship's original complement of 3"/50 dual-purpose secondary battery guns were also replaced with 5"/25 dual-purpose secondary battery guns. The battleship did retain eight of its 5"/51 secondary battery guns, in casemates amidships, four on either side of the superstructure. (*National Archives*)

The USS *Mississippi* went through an updating process in 1942, as had its sister ship, the *New Mexico*. The most prominent external feature of this work was the addition of a large number of small-calibre anti-aircraft guns, visible in this overhead picture of the *Mississippi* taken in 1944. The 5"/51 secondary battery guns normally mounted amidships in armoured casemates seem to be missing in this photograph, no doubt reflecting the US navy's conclusion that they were obsolete. (*National Archives*)

The third battleship in the *New Mexico* class was the USS *Idaho* (BB-42), built by the New York Shipbuilding Corporation, New Jersey. Like the other battleships in its class, it was commissioned too late to see service in the First World War. Between 1931 and 1934 the battleship went through the same modernization process as its sister ships with a new superstructure tower in place of the former cage masts, as shown here. (*National Archives*)

Pictured is the USS *Idaho* entering a floating dry dock on 20 August 1944. The following month it sailed for the West Coast of the United States for upgrading. It was at this point that its 5"/25 dual-purpose secondary battery guns were removed and replaced with eight single 5"/38 dual-purpose secondary battery guns, four on either side of the superstructure amidships. The battleship had its 5"/51 secondary battery guns removed shortly after the Japanese attack at Pearl Harbor. The *Idaho* was the only US navy super dreadnought battleship ever fitted with single 5"/38 dual-purpose secondary battery guns. (*National Archives*)

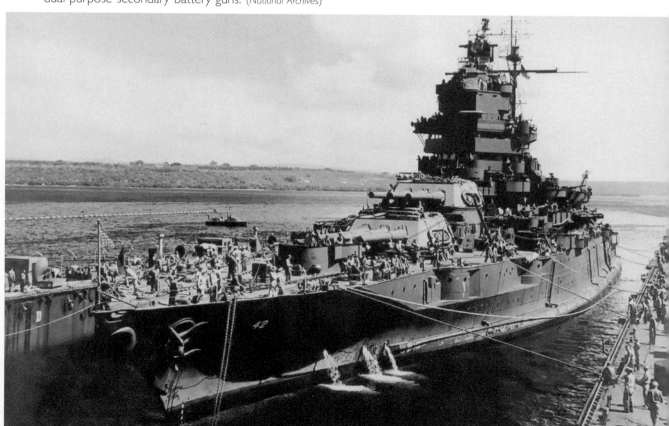

Other than being fitted to the USS *Idaho*, the single 5″/38 calibre dual-purpose secondary battery guns mount was often seen on early US navy destroyers. Pictured is the USS *Kearny* (DD-432) armed with single 5″/38 calibre dual-purpose main battery guns. Besides destroyers, the weapon also appeared on US navy destroyer escorts and eventually fleet auxiliaries such as repair ships. (*National Archives*)

Besides the addition of eight Mk 30 enclosed single 5″/38 secondary dual-purpose guns in 1944, the USS *Idaho*, like its sister ships, received a large number of small-calibre anti-aircraft guns. Pictured on the *Idaho*, manned by the ship's Marine Corps detachment, are three 20mm Single Mount Mk 4 guns of the ship's anti-aircraft battery. The weapon could be elevated from minus 5 degrees to a maximum elevation of plus 87 degrees. (*National Archives*)

Taken in 1944 is this overhead picture of the stern of the USS *Idaho* in a floating dry dock known as the Advance Base Sectional Docks (ABSDs). These docks were built in sections and once towed into place, the sections were welded together based on how large a ship was intended to be placed in them. The largest version, intended for battleships, consisted of ten sections welded together. Each individual section was fitted with two cranes that ran on rails along the top of the dock sections. *(National Archives)*

(Opposite) The two *Tennessee*-class battleships (in their original configuration) had a taller superstructure than previous super dreadnought-type battleships, which surrounded the lower portion of the forward cage mast as seen in this overhead photograph. The picture also shows another external difference of the two *Tennessee*-class battleships from their predecessors: an open-topped flying bridge that surrounds a taller than normal armoured conning tower. *(National Archives)*

(*Above*) The USS *Tennessee* (BB-43) shown here was built at the New York Navy Yard, Brooklyn and was commissioned too late to see service in the First World War. It and its sister ship, the USS *California* (BB-44), were very similar to the preceding three battleships of the *New Mexico* class. They differed in their original external appearance from their predecessors with two stacks instead of one and not being fitted with hull casemated 5"/51 secondary battery guns. They did retain the amidships casemated 5"/51 secondary battery guns. (*National Archives*)

(*Above*) Both *Tennessee*-class battleships were berthed at Pearl Harbor on 7 December 1941 when the Japanese struck. The USS *Tennessee* was berthed inboard of the USS *West Virginia* (BB-48) and therefore escaped any serious damage from the attack. The battleship then went through two upgrade programmes: one in 1942 and the other in 1943. In the 1942 upgrade the ship retained its cage mast but lost its main mast. In the 1943 upgrade the forward cage mast disappeared and the two stacks were faired into a single stack at the rear of the new tower superstructure, as shown in this picture. (*National Archives*)

(*Opposite*) Seen here following its 1943 upgrade is the USS *Tennessee*. During that upgrade it received eight 5"/38 Twin Mount 32 dual-purpose secondary guns, four on either side of the superstructure, in place of the former amidships casemated 5"/51 secondary battery guns. It was also fitted with a large number of small-calibre anti-aircraft guns. Unseen enhancements were improvements to the ship's underwater protection from torpedo strikes. (*National Archives*)

(*Below*) The USS *California* was the sister ship of the USS *Tennessee* and was built at the Mare Island Navy Yard, Vallejo, California. It was the last US navy battleship built on the West Coast of the United States. From August 1921 until December 1922 the USS *California* was the flagship of the US navy's Pacific Fleet. The battleship was modernized between 1929 and 1930. Unlike the US navy super dreadnought battleships commissioned before it, the *California* spent its entire service life in the Pacific, except for one visit to New York City for the opening of the 1939 World's Fair. (*National Archives*)

In December 1922 a General Order was issued that renamed the former Pacific and Atlantic fleets as the 'United States Fleet'. The Pacific Fleet was then referred to as the 'Battle Fleet' and the smaller Atlantic Fleet became the 'Scouting Fleet'. In 1930 the Battle Fleet was renamed the 'Battle Force' and comprised the majority of the US navy's newer battleships. The USS *California*, shown here, served as the flagship of both the Battle Fleet and Battle Force until February 1941. In February 1941 another General Order reorganized the United States Fleet into three fleets: the Pacific Fleet, the Atlantic Fleet and the Asiatic Fleet. (*National Archives*)

From the middle of the 1920s until the summer of 1940 the USS *California* was based at San Pedro, California, until the bulk of the Battle Force was transferred to the naval base at Pearl Harbor. The battleship was berthed at Pearl Harbor's Battleship Row when the Japanese struck on 7 December 1941. A Japanese aerial-delivered bomb and two torpedoes impacted the ship, causing it to list to its port side as seen in this picture taken during the Japanese attack. Casualties on the battleship numbered 100 killed with another 62 wounded. (*National Archives*)

Three days after the Japanese attack on Pearl Harbor the damage inflicted on the USS *California* caused it to sink at its berth up to its superstructure, as seen in this photograph. By March 1942 the ship was refloated and in June 1942 sailed to the Puget Sound Naval Shipyard, Bremerton, Washington for repairs and upgrading. The latter included a new tower superstructure, a single stack faired into the superstructure and a battery of eight amidships 5"/38 dual-purpose secondary battery guns. It took until January 1944 before the battleship was finished. (*National Archives*)

Installed in the Advance Base Sectional Docks (ABSDs) is the USS *California* on 24 August 1944. The battleship took part in the shore bombardment of Saipan in May 1944 and suffered a hit from a Japanese shore battery on 14 June 1944, which killed one and wounded nine. Thereafter the ship performed shore bombardment duties for both the Guam and Tinian landings that took place between July and August 1944. In August 1944 the battleship was placed in dry dock to make repairs to the damage incurred in a collision between it and another US navy battleship. (*National Archives*)

Pictured is the USS *Colorado* (BB-45) in the 1930s, the lead ship in its class. It was built at the New York Shipbuilding Corporation, New Jersey. The big difference between the *Colorado*-class battleships and those super dreadnought classes that came before it was the replacement of the twelve 14-inch guns divided between four main gun battery turrets with eight 16-inch guns divided between the same turrets. (*National Archives*)

The USS *Colorado* had just begun a major rebuild at the Puget Sound Navy Yard, Washington, when the Japanese struck Pearl Harbor. Work on the battleship that had begun in July 1941 was halted and the ship rushed back into fleet service by March 1942 without any major external differences. This meant it still retained its two cage masts during wartime, as shown in this 1943 photograph of the stern of the ship during anti-aircraft gunnery practice, with the main mast in the background. (*National Archives*)

The USS *Colorado* returned for upgrading to the West Coast of the United States in 1944, at which point it lost its main mast, this being replaced by a simple pole mast as seen in this photograph. Rather than the 5"/38 Twin Mount 32 dual-purpose secondary guns, the battleship was fitted with eight 5"/25 dual-purpose secondary guns amidships, four on either side of the superstructure. These weapons were mounted on top of the amidships structure that housed the casemated 5"/51 secondary battery guns. (*National Archives*)

(*Above*) On 24 July 1944, during the bombardment of the Japanese-occupied island of Tinian, the USS *Colorado*, shown here, was struck by twenty-two projectiles fired by enemy shore batteries. Despite the damage inflicted, the ship continued bombardment of its assigned targets. Visible in this photograph are the holes in the battleship's hull from enemy fire. (*National Archives*)

(*Opposite*) Constructed by the Newport News Shipbuilding Company is the USS *Maryland* (BB-46) shown here. It was the second ship in the *Colorado* class of battleships. However, due to an arms control treaty agreed by the United States government, the third battleship in the class, the USS *Washington* (BB-47), was cancelled before being finished in 1922. The USS *Maryland* was overhauled between 1928 and 1929 and had its eight 3"/50 secondary battery dual-purpose guns replaced with eight 5"/25 secondary battery dual-purpose guns. (*National Archives*)

(*Below*) The USS *Maryland* was berthed inboard of the USS *Oklahoma* (BB-37) at Pearl Harbor on 7 December 1941. As a result, the *Oklahoma* took the brunt of the Japanese aerial onslaught. The *Maryland* suffered only two bomb hits and lost four men. The battleship later sailed to the West Coast of the United States for repairs and upgrading, returning to service in February 1942. The ship was struck in the bow by a Japanese aerial-delivered torpedo on 22 June 1944 and the damage sustained is visible in this picture of the ship docked at Pearl Harbor in July 1944. (*National Archives*)

On 29 November 1944 a kamikaze struck the forecastle deck of the USS *Maryland* between main gun turrets Nos 1 and 2, causing extensive damage and killing thirty-one men and wounding another thirty. The battleship returned to Pearl Harbor for repairs and upgrading in December 1944. Pictured is one of the ship's eight 5"/25 dual-purpose secondary battery guns engaged in shore bombardment duty in 1943. The weapon and its crew are protected by a splinter shield. (*National Archives*)

The USS *West Virginia* (BB-48) was constructed at the Newport News Shipbuilding Company. It was berthed outboard of the USS *Tennessee* (BB-43) at Pearl Harbor when the Japanese aerial attack began. The ship took at least seven torpedo hits and two bombs, causing the battleship to sink on an even keel almost to its superstructure, as seen in this picture taken after the attack. (*National Archives*)

Refloated on 17 May 1942, the USS *West Virginia* was stripped of its two cage masts, as seen here, and sailed to the Puget Sound Navy Yard, Washington for repair and rebuilding which took until July 1944. The ship did not reach the Pacific theatre of operations until 18 October 1944 when it engaged in shore bombardment duties in the Philippines. On 21 October 1944 the battleship scraped the ocean floor in an uncharted area and damaged four of its screws. Operating at reduced speed thereafter, the ship continued with its fire support missions. (*National Archives*)

An aerial view of the rebuilt USS *West Virginia* in 1945 shows the battleship fitted with eight of the 5"/38 Twin Mount 32 dual-purpose secondary guns amidships, four on either side of the superstructure. The ship's new superstructure tower mirrored the two *Tennessee*-class battleships when rebuilt. In addition the two stacks were reduced to one, as also occurred with the two *Tennessee*-class ships, and faired into the back of the new superstructure tower. (*National Archives*)

Chapter Four

Fast Battleships

In the years before the expiration of the Washington Naval Treaty in 1936 the US navy had time to contemplate what it wanted to see in its next class of battleships. There were, however, a number of constraints that had to be taken into account by the US navy that influenced the final design of any new class of battleships. These included the width of the Panama Canal, which limited all ships to a beam of less than 110 feet. There were also the treaty requirements agreed to by the US government stating that any new battleships built could not have a standard displacement of more than 35,000 tons and a main battery armament consisting of guns no larger than 14 inches in bore.

Originally the US navy considered merely building an improved version of its last super dreadnought battleships, commissioned in 1923. Wiser heads prevailed, and the US navy decided in 1935 to look at what other countries' navies had built or were having built before committing itself to a design. Based on faulty information, the US navy was led to believe that the Japanese navy's four *Kongo*-class battleships (armed with eight 14-inch main battery guns) had a maximum speed of 26 knots. In reality, they had a maximum speed of 30 knots after their rebuilding in the 1930s. The maximum speed of the US navy's commissioned super dreadnought battleships was only 21 knots.

The US navy's new aircraft carriers (themselves capable of a maximum speed of 33 knots) needed protection from the Japanese navy *Kongo*-class battleships should a conflict ever arise. In response, the US navy proposed a new class of 35,000-ton standard displacement 'fast battleships', capable of a maximum speed of 30 knots and armed with nine 14-inch main battery guns divided between three centreline turrets, each armed with three guns. These new fast battleships would be able to deal with the Japanese *Kongo*-class battleships or any hostile cruisers acting as commerce raiders.

The maximum speed requirements for a new class of fast battleships proved to be a hotly-debated subject within factions of the US navy. Some felt that a maximum speed of only 23 knots was required, while others strongly believed that the new class of fast battleships had to have a maximum speed of 33 knots.

Japan pulled out of the Washington Naval Treaty in 1934, as other signatories to the treaty openly expressed their desires to build ever-faster battleships once the treaty requirements officially lapsed at the end of 1936. The US navy went in the other direction and proposed a slightly slower battleship class with a maximum speed of 27 knots, armed with twelve 14-inch main battery guns divided between three centreline turrets, each armed with four guns.

This proposed new battleship class reflected the US navy's emphasis on armour protection and armament over speed, as the weight of the main battery guns determined the amount of armour protection that could be devoted to a battleship design and not exceed the 35,000-ton standard displacement limit. Following this train of thought, the US navy then proposed another design go-around: an up-armoured 27-knot battleship class armed with twelve 12-inch guns divided between three centreline turrets, each armed with four guns.

Changing Their Minds

In 1937 the US navy began rethinking what it wanted in a new class of battleships. Rather than placing the number one priority on armour protection, they decided to prioritize firepower, followed by speed and armour protection. This resulted in the US navy now proposing a 35,000-ton standard load displacement battleship armed with nine 16-inch main battery guns divided between three centreline turrets, each armed with three guns, two forward of the ship's superstructure and one aft.

The US navy's jump to proposing 16-inch main battery guns on a new class of fast battleships came about because of an escalator clause in the London Naval Disarmament Treaty of 1936. That clause allowed a signatory to the various naval arms control treaties to arm their new battleship classes with 16-inch main battery guns if a non-signatory nation (Japan in this case) failed to comply with the treaty limits that had set 14-inch main battery guns as the largest that could be mounted on battleships.

President Franklin D. Roosevelt had not been in favour of mounting 16-inch guns on the US navy's new class of proposed battleships. However, he could not talk the Japanese government into agreeing to restrict mounting guns no larger than 14 inches on their battleships. Roosevelt therefore reluctantly approved the US navy's plan for mounting 16-inch guns on their newest battleship class.

Traditionally, battleship armour is supposed to be thick enough to withstand the projectiles of its own main battery guns at normal combat ranges. However, with treaty limitations forcing the US navy to keep its next class of battleships under 35,000 standard displacement tons, a decision was made that its armour would only be proof against 14-inch main battery guns, rather than its own 16-inch main battery guns.

In great secrecy, the Japanese navy had embarked on building its *Yamato*-class battleships. They had a full load displacement of 69,000 tons and a maximum speed of 27 knots. Armament consisted of nine 18.1-inch main battery guns divided between three centreline turrets, each armed with three guns. So well did the Japanese navy hide the fact from the world that its newest class of battleships was armed with such large main battery guns that the US navy firmly believed that its proposed new 16-inch main battery gun-armed battleships would be sufficient in any future conflict with Japan. However, plans were laid for 18-inch main battery guns on a future class of US navy battleships if the need arose.

North Carolina Class

After much debate on what the next class of US navy battleships would look like, the process eventually resulted in the authorization by Congress in 1937 of the two *North Carolina* fast battleships: the USS *North Carolina* (BB-55) and the USS *Washington* (BB-56). Both were commissioned in 1941. They had a full load displacement that ranged from 41,000 tons to 44,800 tons, and a crew complement that included 2,500 officers and men during wartime. The large increase in crew compared to previous battleships was primarily due to the large number of anti-aircraft guns that needed manning.

The two *North Carolina* fast battleships were 728 feet 9 inches long and had a beam of 108 feet 4 inches. Strangely enough, when the designs for the two *North Carolina* fast battleships were finalized by the US navy, their projected role as escorts for aircraft carriers was rejected. Rather, their main purpose was to find, engage and destroy their Japanese navy counterparts in a decisive battle that would determine the outcome of the war between the two nations.

The *North Carolina* fast battleships mounted nine 16-inch main battery guns divided between three centreline turrets, each with three guns, two turrets forward of the ship's superstructure and one aft. The secondary gun battery on the ships consisted of twenty dual-purpose 5-inch guns divided between ten power-operated armoured mounts armed with two guns each, with five of these mounts on either side of the ship's superstructure.

The 5-inch secondary battery guns on the *North Carolina* fast battleships were referred to in official documents as the 5"/38cal and had a barrel length of 190 inches (15 feet 10 inches). They were designated as dual-purpose guns because their maximum elevation of 85 degrees meant they could engage both surface and aerial targets. Other navies kept the inefficient arrangement of separate secondary and anti-aircraft guns. Numerous additional smaller-calibre anti-aircraft guns would become standard on the *North Carolina* fast battleships, as they would on the super dreadnoughts that came before them during the Second World War.

The sloping vertical armoured faces on the ship's main battery turrets of the *North Carolina*-class fast battleships were 16 inches thick. The upper hull armoured belts on the ships had a maximum thickness of 12 inches and were inclined inwards at 15 degrees to provide protection from long-range high-trajectory plunging fire.

Behind this inclined hull armour belt was an additional armour backing plate. The conning tower was built from armour 14.7 inches thick at its maximum. The ship's three armoured hull decks had a combined thickness of 7.07 inches. Aerial-delivered ordnance was clearly a major threat to all battleships and the amount of horizontal armour also reflected this threat. Besides an internal torpedo belt, the ships had a triple bottom. This was in lieu of the double bottom seen on all previous US navy battleship classes.

Unlike the super dreadnought battleships that came before them, the *North Carolina*-class fast battleships and those classes that followed lacked the heavy-duty tripod masts for mounting the ship's various fire-control instruments. Instead there was a massive centreline superstructure bridge assembly near the front of the ship, aft of the forward main gun battery turrets. Within the bridge on the *North Carolina*-class fast battleships was a heavily-armoured conning tower. In actual wartime practice, most US navy battleship officers preferred the better visibility of the unarmoured navigation bridge when engaged in combat to the more limited visibility of the armoured conning towers.

Mounted on the top of the *North Carolina*'s superstructure bridge assembly were air-search and fire-control radars. Within their hulls were the mechanical/electric fire-control analogue computers that processed all the information supplied by the various optical and radar-guided sensors of the ships and came up with fire-control solutions that maximized the effectiveness of the ship's main guns and secondary battery guns in a variety of environmental conditions.

The radar and other electronic systems on the *North Carolina*-class fast battleships and the follow-on fast battleships of the US navy were far in advance of what was installed on their Japanese navy counterparts and provided them and other US navy ships with an unmatched technical superiority during the latter part of the Second World War.

The *North Carolina*-class fast battleships had geared steam turbine engines rather than the turboelectric engines of the last few classes of the US navy's super dreadnoughts. The decision to go with geared steam turbine engines (powered by fuel oil-fired boilers) on the *North Carolina*-class fast battleships was due to their lighter weight than the turboelectric engine counterparts. This was an important consideration as the ship designers had to balance the various components of the ships and make them all fit within arms control treaties limitations. The maximum speed of the *North Carolina*-class fast battleships was 28 knots. All of the follow-on classes of US navy fast battleships would have geared steam turbine engines, powered by fuel oil-fired boilers.

South Dakota Class

Following the *North Carolina*-class fast battleships Congress authorized two additional fast battleships, later raised to four, of a new *South Dakota* class. This included the

USS *South Dakota* (BB-57), USS *Indiana* (BB-58), USS *Massachusetts* (BB-59) and USS *Alabama* (BB-60). All four ships would be commissioned in 1942 and retain the same main gun battery arrangement as their direct predecessors.

The *South Dakota* class of fast battleships was intentionally made 68 feet shorter than the preceding *North Carolina* class. This meant less space to armour. They had thick enough armour to withstand hits from their own 16-inch main battery guns at normal combat ranges, whereas the two *North Carolina*-class fast battleships that came before them only had protection from 14-inch main battery guns.

The armour on the sloping vertical face of the *South Dakota*-class fast battleships' main gun battery turrets was 16 inches thick. The hull belt armour had a maximum thickness of 12.2 inches, with a backing plate behind it. Conning tower armour was 16 inches at its maximum. At its uppermost portion the hull belt armour was inclined at 19 degrees inward. The lower portion of the hull belt armour, which extended to the bottom of the ship's hull, was inclined outward from the keel to protect the hull under the waterline from long-range high-trajectory plunging fire. As with the *North Carolina* class, the *South Dakota*-class fast battleships had three horizontal armoured decks, as well as a triple bottom and a torpedo belt.

The new *South Dakota*-class fast battleships had a full load displacement that ranged from 42,000 tons to 45,200 tons and a crew complement that included 2,500 officers and men during wartime. The ships were 680 feet long and had a beam of 108 feet 2 inches. All four *South Dakota* fast battleships had a maximum speed of 27 knots; one knot slower than the *North Carolina*-class fast battleships.

Looking very similar to the *North Carolina* class, the *South Dakota* class of fast battleships can be distinguished by the fact that they have a single stack just behind the superstructure bridge assembly in comparison to the two seen on their direct predecessors. A 'stack' is a pipe-like structure housing the uptakes that convey smoke and gases from a ship's boilers.

A distinguishing feature of the USS *South Dakota* (BB-57) was the fact that it had only sixteen 5-inch dual-purpose guns divided between eight power-operated armoured mounts of two guns each, four on either side of the ship's superstructure, on a single deck. This was done to make room for staff spaces. The other three ships of the *South Dakota* class – USS *Indiana* (BB-58), USS *Massachusetts* (BB-59) and USS *Alabama* (BB-60) – all had the same secondary gun battery arrangement seen on the *North Carolina* class, with twenty dual-purpose 5-inch guns divided between ten power-operated armoured mounts, each armed with two guns. There were five of these two-gun 5-inch powered armoured mounts on either side of the ship's super-structure located on two superimposed decks.

The two *North Carolina*-class and four *South Dakota*-class fast battleships all per-formed yeoman service during the Second World War. Not in the original role envisioned for them as being in the forefront of the US navy Pacific Fleet carrying the battle to the enemy – that role was inherited by the aircraft carrier – but in a supporting role as aircraft carrier escorts. They would also go on to provide valuable fire support to American soldiers and Marines going ashore on enemy-occupied possessions during amphibious landings. Of the original six US navy fast battleships, three have been preserved as museum ships: the USS *North Carolina* (BB-55), USS *Massachusetts* (BB-59) and USS *Alabama* (BB-60). The other three ships went through a final decommissioning by 1947 and were eventually scrapped.

The only engagement that occurred between the first six US navy fast battleships and their Japanese navy counterparts took place on the night of 14 November 1942. The Japanese navy battleship *Kirishima* (one of the four *Kongo*-class battleships) identified and then opened fire on the USS *South Dakota* (BB-57) which could not return fire due to an electrical power malfunction. At the same time, the Japanese sailors failed to spot the USS *Washington* (BB-56). The *Washington*'s fire-control radar allowed it to engage and fatally cripple the battleship *Kirishima* in a span of five and a half minutes with its main and secondary gun batteries. The Japanese battleship sank on the following day.

Iowa Class

There were six fast battleships authorized by Congress following the four *South Dakota*-class ships. These would be the *Iowa*-class fast battleships, comprising the USS *Iowa* (BB-61), the USS *New Jersey* (BB-62), the USS *Missouri* (BB-63), the USS *Wisconsin* (BB-64), the USS *Illinois* (BB-65) and the USS *Kentucky* (BB-66). The first four ships of the *Iowa* class were commissioned between 1943 and 1944, while the last two would not be completed in time to see service during the Second World War and were eventually scrapped without ever being completed.

The *Iowa*-class fast battleships were lengthened *South Dakota*-class variants with the same general main and secondary battery armament arrangement as the last three *South Dakota*-class battleships with nine 16-inch main battery guns and twenty 5-inch guns divided between ten twin-gun powered armoured mounts. Like the *South Dakota*-class ships, they were proof against their own 16-inch main gun battery fire at normal combat ranges. These were the first modern US navy battleships designed without being influenced by treaty restrictions.

The 16-inch main battery guns on the *Iowa*-class fast battleships were listed in US navy documents as 16in/50cal, meaning they were 66 feet 7 inches long. This extra length increased their range and penetration power. The calibre refers to the length of the barrel relative to the size of the gun, so 16in/50 meant the barrel was 50 times 16 inches measured from the breech. The 16-inch main battery guns on the US navy's preceding *North Carolina*- and *South Dakota*-class fast battleships were listed in US navy documents as 16in/45cal, meaning they were 60 feet long.

There had been some initial consideration for mounting 18-inch guns on the *Iowa*-class fast battleships but due to their size only six could have been mounted and the US navy decided to stay with nine 16-inch guns.

The armour arrangement on the *Iowa* class was generally modelled on the preceding *South Dakota*-class fast battleships but was thicker in some areas. The sloping vertical face of the *Iowa*-class fast battleships' main gun battery turrets was 18 inches thick and the conning tower armour was 17.5 inches at its maximum. As with the *South Dakota* class, the *Iowa*-class fast battleships had three horizontal armoured decks as well as a triple bottom and a torpedo belt.

The *Iowa*-class fast battleships had a Second World War standard load displacement of 45,000 tons. This became possible under the escalator clause in the Second London Treaty when the US navy became aware of the Japanese navy and its building of the *Yamato*-class battleships during the Second World War. The full load displacement of the *Iowa* class of fast battleships eventually rose to 57,600 tons.

The *Iowa*-class ships had a length between 887 feet 3 inches and 887 feet 7 inches. Their beam ranged between 108 feet 1 inch and 108 feet 3 inches. During the Second World War there was a complement of 2,700 officers and men. Maximum speed of the *Iowa*-class fast battleships was 33 knots.

Unlike the *South Dakota* class that had a single stack behind the superstructure bridge assembly, the *Iowa*-class fast battleship reverted to a two-stack arrangement with the forward stack faired into the rear of the superstructure bridge assembly. This was necessary as its higher speed design required more boilers than a single stack could handle.

Like the *North Carolina* and *South Dakota* classes that came before them, the *Iowa*-class fast battleships were not intended to escort aircraft carriers; rather they were to be the 'tip of the spear' of the US navy. It is a testament to their design that when the aircraft carrier replaced them in that role they still provided useful service in defending the carriers from attack and contributed mightily to the eventual victory in the Pacific Ocean theatre of operations. The US navy chose the main deck of the USS *Missouri* (BB-63) on 2 September 1945 for the Japanese government to sign the surrender documents that brought an end to the Second World War.

Unlike the *North Carolina*- and *South Dakota*-class fast battleships that were quickly pulled from service following the Second World War, the *Iowa*-class ships would become the longest-serving battleships in the US navy with the last one going through its final decommissioning in 1992. All four *Iowa*-class fast battleships would go on to see second lives as museum ships.

Following the Second World War the *Iowa*-class fast battleships saw service in a number of Cold War conflicts including the Korean and Vietnam Wars, also in the Lebanon and Operation DESERT STORM. They would also be upgraded with the latest in electronics and weapon technology to keep them as viable weapon platforms until the end of the Cold War in 1991. The postwar weapon additions included anti-ship cruise missiles and long-range cruise missiles; depending on the model, able to engage both surface and land targets. To protect the *Iowa*-class fast battleships from enemy anti-ship cruise missiles, they were eventually fitted with the 20mm Phalanx close-in weapon system (CIWS).

Montana Class

There were at one point plans to build a follow-on class of battleships even larger and better armed than the *Iowa*-class fast battleships. This would have been the *Montana*-class battleships scheduled to include the USS *Montana* (BB-67), USS *Ohio* (BB-68), USS *Maine* (BB-69) and USS *New Hampshire* (BB-70). All four ships were authorized in 1940 but none of the keels for these ships was ever laid and all were cancelled by 1943.

The *Montana* class of battleships was to have a full load displacement, if completed, of almost 71,000 tons and be 921 feet long. With a planned beam of 121 feet 2 inches they would not be able to use the Panama Canal. Maximum speed was to be 28 knots. Their main armament was to be twelve 16-inch guns divided between four turrets each armed with three guns, two forward and two aft of the superstructure. These would be the same 16in/50 guns mounted on the preceding *Iowa*-class battleships.

Shown at its launching on 13 June 1940 is the USS *North Carolina* (BB-55), the first of the US navy's 'fast battleships'. It was constructed at the New York Navy Yard, Brooklyn. The ship was the US navy's first new battleship since the commissioning of the USS *West Virginia* (BB-48) in December 1923. The *North Carolina* had a maximum trial speed of 27 knots, 6 knots faster than all previous classes of US navy super dreadnought-type battleships. This increase in speed was intended to match what was then considered the maximum speed of their Japanese navy counterparts.
(*National Archives*)

(*Above*) Near the very top of the USS *North Carolina* foremast is the armoured traversable Mk 38 main battery director. It incorporated a 26.5-foot stereoscopic range-finder with its windows projecting out either side of the director's crew compartment. There was a second Mk 38 main gun battery director located at the rear of the ship's superstructure, known as the 'after main battery director'. Visible on the roof of the navigation bridge is one of the battleship's four Mk 37 directors for the secondary gun battery. (*National Archives*)

(*Above*) Visible in this postwar photograph of the USS *North Carolina* are the armoured stereoscopic range-finder enclosures for both forward main gun turrets. The Mk 37 director located just above the navigation bridge is missing its antenna. In case of damage to either of the battleship's two Mk 38 main battery directors, any of the ship's four Mk 37 secondary battery directors could take over control of the main gun battery turrets, albeit with lesser capability. Each individual main gun turret could also operate on its own, independent of the Mk 38 main gun battery or if the Mk 37 secondary battery directors were rendered inoperable. (*National Archives*)

(*Opposite*) The US navy began including stereoscopic range-finders on the main gun battery turrets of its fast battleships in 1940. Pictured is a rear view of the large, projecting armoured enclosure for the starboard side of the No. 1 main gun battery turret's stereoscopic range-finder. This picture was taken on the preserved USS *North Carolina* berthed at Wilmington, North Carolina. The two smaller armoured enclosures seen at the front of the turret are for telescopes. (*Paul Hannah*)

(*Above*) The second battleship in the *North Carolina* class was the USS *Washington* (BB-56) shown here in 1942. Unlike the US navy's dreadnought and super dreadnought battleships that all had 'broken decks', as the portion of the stern upon which the aft main gun battery turrets were mounted was one deck lower than the forecastle deck, the *North Carolina*-class battleships and all the fast battleships that followed had 'flush decks' as they were unbroken from bow to stern. This also meant that the battleship's flush deck was now considered the 'main deck' or uppermost 'complete deck'. (*National Archives*)

(*Opposite, above*) A wartime image of the USS *Washington* showing the crew's laundry hung out to dry on the ship's life-lines. The USS *Washington* was built at the Philadelphia Navy Yard, Philadelphia. The *Washington* and its sister ship, USS *North Carolina* were plagued early on by serious vibration issues that prevented the battleships from reaching their maximum speed. This problem was soon resolved by changing the configuration of the ships' four propellers and adding extra bracing to the ships' propulsion machinery. (*National Archives*)

(*Opposite, below*) Pictured in April 1942 at Scapa Flow, Scotland is the USS *Washington* in the foreground and the aircraft carrier the USS *Wasp* (CV-7) in the background. The battleship was transferred to the Pacific theatre of operations with Pacific Fleet Battleship Division 6 in August 1942. The ship took part in the Battle of Guadalcanal in November 1942 and sank the Japanese navy battleship IJN *Kirishima*. It spent most of the remainder of the Second World War with the Pacific Fleet, either escorting aircraft carriers or engaging in shore bombardment duties. (*National Archives*)

The USS *South Dakota* (BB-57) is shown here on 9 August 1943, serving with the Royal Navy's Home Fleet. The battleship was the first commissioned in the four-ship *South Dakota* class and was built at the New York Shipbuilding Corporation, New Jersey. The *South Dakota*-class fast battleships can be readily identified by having only one stack faired into the rear of the superstructure tower, in contrast to the *North Carolina*-class battleship's two-stack arrangement. (*National Archives*)

This picture shows one of the engine rooms of the USS *South Dakota* during the Second World War. The sailor is looking at the operating face of the ship's high-pressure fuel oil-fired burners. During the Battle of Santa Cruz Island in late 1942 an enemy 500-lb aerial-delivered bomb struck the battleship's turret No. 1 without causing any major damage. However, the *South Dakota* was credited with downing twenty-six Japanese aircraft in return. During the November 1942 Battle of Guadalcanal the battleship suffered forty-two hits from three enemy ships. (*National Archives*)

A stern view of the USS *Indiana* (BB-58) taken in 1942. Visible on the ship's fantail are three observation floatplanes. Also seen in this photograph is the Mk 38 after main battery director, just behind the battleship's single stack. Constructed by the Newport News Shipbuilding Company, the ship was the second in the *South Dakota* class. The weather or main deck of all the fast battleships was referred to as the 'bomb deck', with the second deck being the main armour deck for the ship. Just below the main armour deck was a splinter deck the length of the armoured citadel. (*National Archives*)

Visible on the fantail of the USS *Indiana* is a group of four 20mm Single Mount Mk 4 anti-aircraft guns protected by their own individual splinter shields, as well as a larger, waist-high splinter shield mounted in front of them. In the background are two of the 5"/38 Twin Mount 32 turrets, as well as a number of small-calibre anti-aircraft guns. The ship's Mk 37 secondary battery directors could be employed to control the 40mm anti-aircraft guns if the need arose. (*National Archives*)

Shown is the USS *Indiana* in September 1942. The battleship spent its entire service career in the Pacific theatre of operations, either escorting aircraft carriers or on shore bombardment duties. During the June 1944 Battle of the Philippine Sea, the *Indiana* accounted for five enemy aircraft and dodged an enemy aerial-delivered torpedo. In that same engagement a Japanese torpedo plane struck the starboard side of the battleship's hull but bounced off with only minor damage inflicted. The battleship was decommissioned in September 1947 and sold for scrapping in September 1963. (*National Archives*)

The USS *Massachusetts* (BB-59) pictured here was built by the Bethlehem Steel Company, Quincy, Massachusetts. It was the third ship commissioned in the *South Dakota* class. During the 1942 Anglo-American invasion of French North Africa the battleship engaged in combat with the incomplete and berthed Vichy French navy battleship, the *Jean Bart*. In the ensuing engagement the American battleship put out of action the French battleship's one operational main gun battery turret. (*National Archives*)

An aerial view of the USS *Massachusetts*, a *South Dakota*-class battleship, dated 1944. Due to the US navy's desire to up-armour the *South Dakota*-class battleships compared to the earlier two-ship *North Carolina* class without exceeding a 35,000-ton displacement limit, they shortened the length of the armoured citadel that contained the battleship's most vulnerable parts. In turn, this forced the ship's designers to shorten the waterline length of the *South Dakota* class by 47 feet compared to the earlier *North Carolina* class. (*National Archives*)

The USS *Alabama* (BB-60) shown here prior to entering combat in late 1942 or early 1943 was the last vessel commissioned in the four-battleship *South Dakota* class. It was constructed at the Norfolk Navy Yard, Portsmouth, Virginia. The lateral metal pole which extends out from the foremast, mounted just behind the tower superstructure, is referred to in naval terms as the 'yard'. The uppermost portion of the tower mast is known as the 'truck'. The pennants attached to the yard are signal flags intended to be used in daytime in place of radios. (*National Archives*)

A portside view of the USS *Alabama* during the Second World War, without its observation floatplanes on the ship's fantail. The reduced length of the *South Dakota*-class battleships resulted in them being both more manoeuvrable and more stable as firing platforms than the preceding *North Carolina* class. The battleship retains a small main mast between the stack and the Mk 38 aft main battery director. (*National Archives*)

A 1945 photograph showing the USS *Alabama* with a new camouflage paint scheme. The ship now features a new taller and sturdier main mast, eventually fitted to all the *South Dakota*-class and earlier *North Carolina*-class battleships, which sported an SC-2 air search radar antenna. During the Second World War the US navy employed up to twelve different camouflage paint schemes, labelled Measure 1 through to Measure 33. They were intended to make it difficult for enemy ships, submarines and aircraft to detect the battleships at a distance through optical devices and to obfuscate their speed and heading. (*National Archives*)

Preserved as a museum ship is the USS *Iowa* (BB-61), shown here in 2012 off the coast of Southern California. It was the first ship commissioned in the *Iowa* class of projected six fast battleships that were unhindered by any arms control treaty limitations and therefore larger in size and displacement than any previous US navy battleships. They were the answer to the unknown next generation of Japanese battleships, yet were still able to pass through the locks of the Panama Canal. (*US Navy*)

Visible is the Mk 38 main battery director, located on top of the tower superstructure, of a preserved *Iowa*-class battleship. On the roof of the director is a Mk 13 fire-control radar antenna. The large box-shaped structure directly below the director housed equipment for an electronic countermeasures system. It was applied to the USS *Iowa* and two of its sister ships – the USS *Missouri* (BB-63) and the USS *Wisconsin* (BB-64) – in the 1980s when they were reactivated and modernized for use once again. (*Vladimir Yakubov*)

Parked in a dry dock is the USS *Iowa*. The battleship was constructed by the New York Navy Yard, Brooklyn. Of the six *Iowa*-class battleships originally intended to be built, two were later cancelled: the USS *Illinois* (BB-65) and the USS *Kentucky* (BB-66.) The *Illinois* was cancelled on the conclusion of the Second World War and the *Kentucky* in 1958. Very evident in this photograph is the USS *Iowa*'s long narrow hull that allowed it and its sister-ships a maximum speed of 33 knots. (*US Navy*)

Reflecting the gun and armour race between navies before and during the Second World War, the *Iowa* class was armed with nine larger-calibre 16-inch main battery guns than those mounted on previous US navy battleships. This increase in barrel length in conjunction with an increase in the propellant charge meant that the *Iowa*-class battleship's main battery guns boasted improved range and muzzle velocity. The increase in muzzle velocity meant that the ship's armour-piercing projectiles could penetrate thicker armour plate. Pictured are the Nos 1 and 2 super-firing main gun battery turrets on an *Iowa*-class battleship during a postwar refit. (*US Navy*)

The 16-inch projectiles on the *Iowa*-class battleships were stowed upright, as seen in this picture, and held in place by steel restraints known as 'projectile lashings' on two levels of the barbette of the 16-inch gun armed turrets. The projectiles were transported to the projectile decks by overhead trolley to a point under an open hatch. From there, they were raised or lowered to a shell deck by an electrically-powered whip or hoist. Moving the 16-inch projectiles around the shell decks was done with ropes and pulleys powered by electrically-powered capstans. To aid in the movement of the upright 16-inch projectiles, the shell decks were lightly lubricated. (*US Navy*)

The 16-inch armour-piercing projectiles for the main battery guns on the *Iowa* class weighed 2,700lb. Pictured is a 1,900-lb high-capacity (HC) high-explosive 16-inch projectile on an *Iowa*-class battleship being lowered into the ship's hull by a projectile carrier, which consisted of a steel frame that was adapted for hoisting a large projectile in either horizontal or vertical positions. Due to the inherent danger of loading or unloading ammunition on board ships during the hours of darkness, the US navy bans such activity, except in case of emergency. (*US Navy*)

A display on a preserved US navy *Iowa*-class battleship shows the progression in size of the armour-piercing projectiles employed by US navy battleships over the many decades. From left to right: a 12-inch, a 14-inch and a 16-inch projectile. The USS *Texas* was the first pre-dreadnought-type battleship to mount 12-inch main battery guns, while the USS *New York* was the first dreadnought-type battleship to mount 14-inch main battery guns. The USS *Colorado* was the first super dreadnought to feature 16-inch main battery guns. (*Paul Hannah*)

Once a determination was made on the type of round to be fired from an *Iowa*-class battleship's 16-inch main gun battery turret, the appropriate projectile was moved to the bottom portion of an electric-hydraulic power hoist. Once the projectile was loaded onto the hoist, it travelled to a turret and then slid onto a cradle, as seen in this photograph. At that point the cradle was lowered in line with a spanning tray, seen here in front of the gun captain wearing protective gloves. Once in the spanning tray, a power rammer shoved the projectile into the breech of the gun. (*US Navy*)

At the same time as a 16-inch projectile is moving up from a shell deck to a turret on a US navy fast battleship, silk powder (propellant) bags are placed in an electric-hydraulic-powered powder-handling hoist, as shown here on an unnamed battleship. They are then transported in groups of three to the turret. In the powder magazine, located adjacent to the bottom of a turret barbette, the powder bags are stored in metal tubes in a horizontal position. (*National Archives*)

Upon the powder bags reaching the 16-inch gun turret, they are pulled out of the hoist door (which drops downward) opposite the gun captain, and then rolled onto the visible spanning tray. When there are six powder bags on the spanner tray, the power rammer shoves them into the breech of the gun, behind the projectile. Once the six powder bags are loaded into the gun's breech, the spanning tray will then be returned to an upright position to make room for the gun's recoil travel. (*US Navy*)

A US navy petty officer stands next to the closed breech plug of a 16-inch gun on an *Iowa*-class battleship. Naval guns that did not employ powder bags had a sliding-wedge breech block. US navy battleships armed with 16-inch guns employed an interrupted-screw mechanism to close the gun's breech plug. Because the hinged breech plug on such a large weapon weighed as much as 1,400lb, the gun crew was provided with a powered device to assist them. The two large circular objects on either side of the top of the gun are the aft ends of the counter-recoil cylinders. (*US Navy*)

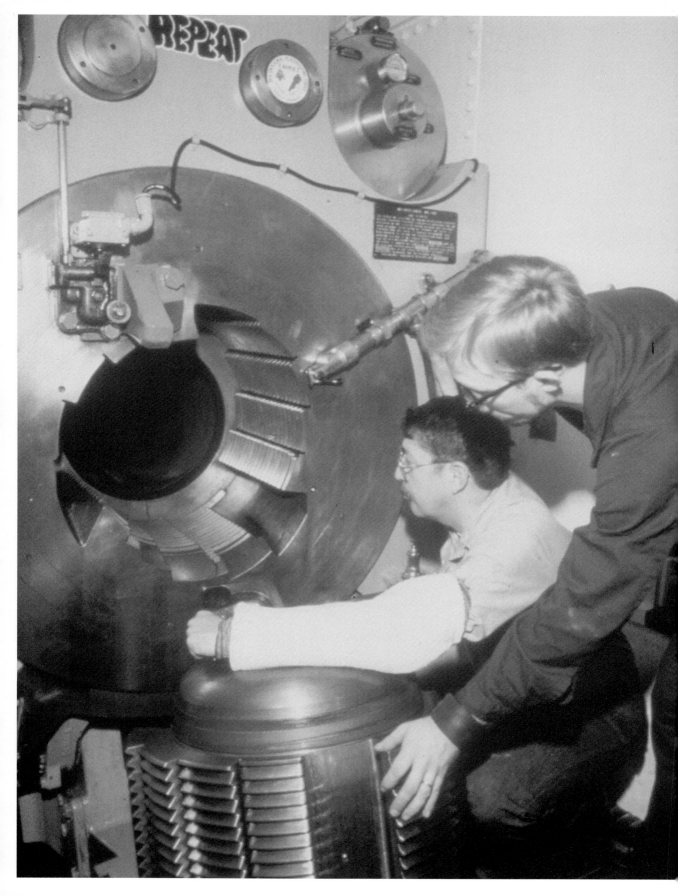

(*Above*) Although the 16-inch main battery gun turrets on the US navy's fast battleships had all the equipment needed to direct and fire their own weapons, referred to as 'local control', the preferred method of firing was by automatic director control, also known as 'remote control'. Directors, such as the Mk 38, transmitted their data to one of the ship's two main battery plotting rooms, one of which is shown here, located within the armoured confines of the ship's citadel where it was corrected by an electro-mechanical computer. The main battery plotting room shared the corrected information with the battleship's combat information centre (CIC), also located within the armoured citadel of the battleship. (*US Navy*)

(*Opposite*) Prior to the closing of the interrupted-screw mechanism seen here on a 16-inch main battery gun, which incorporated an obturation (gas-check) system intended to prevent hot, burning powder gases from spitting out from around the breech plug when the weapon was fired, the primer-man inserts a small primer into the primer chamber of the breech pad. When the gun was fired, the primer detonated one end of a silk powder bag that had sewed into it a quilted ignition pad containing black powder which in turn set off the propellant charge contained within the powder bags. (*US Navy*)

(*Above*) Guns of any size need to be cleaned after being fired. This rule also applied to the 16-inch guns on US navy fast battleships, as is seen in this picture of a junior enlisted man performing the necessary task. The sailor is lying in the gun's 'chamber', the enlarged rear part of the inside of a gun, which is smooth rather than rifled as is the gun's barrel and held the powder bags, located behind the projectile. The projectile, loaded before the powder bags, is pushed far enough into the chamber that a projectile's soft copper rotating bands are engaged with the aft end of the barrel's rifling. In the foreground of the picture is the gun's spanning tray. (*National Archives*)

(*Opposite*) Once an *Iowa*-class battleship's combat information centre (CIC) decided on what target or targets the main gun battery turrets needed to engage based upon the threat they posed, the ship's 'gunnery officer' (or gun boss), who acted as the ship's chief fire-control officer, conferred with the battleship's captain to evaluate the information supplied. It was they who then made the final determination on designating a target or targets. They passed orders to the 'main battery assistant' on when and what to shoot at. Pictured is the USS *Iowa* firing a full salvo from its main battery guns during a training exercise in the 1980s. (*US Navy*)

Fast-moving anti-ship missiles had become an ever-growing threat to US navy ships by the mid-1960s. In response, a number of missile systems were mounted on board US navy ships that were intended to engage and destroy incoming anti-ship missiles at varying ranges. For those anti-ship missiles that managed to avoid destruction at longer ranges and were within 1,500 to 500 yards of a ship, the US navy sought out a point-defence weapon that eventually evolved into the 20mm Phalanx Close-In Weapon System (CIWS) seen here. When the four ships of the *Iowa* class were reactivated in the 1980s they all received four CIWS. (*US Navy*)

Another point-defence weapon that appeared on the reactivated *Iowa*-class battleships in the 1980s was the .50 calibre machine gun seen here. Rather than the water-cooled version that had appeared on pre-Second World War US navy battleships, the version that went back onto the battleships was the air-cooled version, designated M2. The weapon was no longer intended as an anti-aircraft weapon but to protect US navy ships from small, fast, explosive-laden suicide boats when in foreign ports or in confined waterways. (*Paul Hannah*)

During the reactivation of the *Iowa*-class battleships in the 1980s they were fitted with the RGM-84 Harpoon surface-to-surface anti-ship missile system, an example of which is shown being fired from its launcher on board a US navy ship. The 15-foot-long missile delivers a 490-lb high-explosive warhead to its target at a speed of 530 miles per hour. It has a sea-skimming cruise trajectory to minimize its exposure to enemy defensive measures and a range of approximately 60 miles. Seen in the foreground and background of this picture are Mk 36 SRBOC (Super Rapid Blooming Offboard Chaff) launchers that fired cartridges intended to confuse incoming enemy anti-ship missiles. These were fitted onto all the reactivated *Iowa*-class battleships in the 1980s. (*US Navy*)

To increase the combat reach of the reactivated *Iowa*-class battleships in the 1980s, they were fitted with eight large armoured box launchers (ABLs), two of which are seen here, that each housed four BGM-109 series Tomahawk cruise missiles and their respective launching mechanisms. The 20.5-foot-long missiles carried either a conventional or nuclear warhead to a range of up to 1,500 miles. Their first combat employment occurred during Operation DESERT STORM in 1991. (*Vladimir Yakubov*)

This July 1984 overhead photograph of USS *Iowa* firing a salvo from its main gun battery shows the location of the eight armoured box launchers (ABLs) for the BGM-109 series Tomahawk cruise missiles. Four are located in-between the ship's two stacks, two on either side, and the other four are seen on either side of the aft main battery director. Also visible are the four quadruple launchers for the Harpoon missiles on either side of the aft stack. On the fantail of the battleship is a helicopter landing pad, which replaced the launching catapult for observation floatplanes that had been removed in 1949. (*US Navy*)

With the introduction of the newest generation of naval missiles on board the reactivated *Iowa*-class battleships in the 1980s, the original Second World War vintage combat information centre (CIC) was not up to the task of overseeing the employment of such weapons. As a result, some of the interior space within the armoured citadel of the ship's hull was converted into what was referred to as a 'combat engagement centre' (CEC). One of the control consoles for the BGM-109 series Tomahawk cruise missiles is seen here in the CEC on a preserved *Iowa*-class battleship. *(Paul Hannah)*

Visible in this picture of an *Iowa*-class battleship is the launching of one of the ship's BGM-109 series Tomahawk cruise missiles. To provide space for mounting of the ABLs for these missiles on the *Iowa*-class battleships, they had four of their 5"/38 Twin Mount 32s, two on either side of their superstructures, removed. The wooden deck, as seen in this picture, was a standard feature on most US navy battleships as it provided insulation to below-deck spaces and offered safer footing to the crew in wet weather. *(US Navy)*

(*Above*) The second ship in the *Iowa* class was the USS *New Jersey* (BB-62), seen here during the Second World War, bristling with small-calibre anti-aircraft guns. The battleship was built by the Philadelphia Navy Yard, Pennsylvania. The ship first entered into combat at the beginning of 1944 and performed the same roles conducted by the other *Iowa*-class battleships during the Second World War: aircraft carrier protection and shore bombardment. The *Iowa*-class battleships never met their Japanese navy counterparts in battle; the objective for which they had been designed. (*National Archives*)

(*Opposite, above*) In their original configuration the USS *Iowa* and the USS *New Jersey* both featured an open walkway mounted around the outside of their armoured conning towers. This was a design feature that had first appeared on the *South Dakota*-class battleships. During Second World War refits, both battleships eventually had their open walkways enclosed, as shown in this photograph of the preserved USS *New Jersey*. The USS *Missouri* and USS *Wisconsin* were both commissioned with an enclosed walkway around their armoured conning towers. (*Vladimir Yakubov*)

(*Opposite, below*) The armoured conning tower on the *Iowa*-class battleships was 17.3 inches thick, as is evident from this picture of one of the access doors to the conning tower on a preserved *Iowa*-class battleship. The roof of the armoured conning tower was 7.25 inches thick. Both the *South Dakota*- and *Iowa*-class battleships had an internal main armour belt in the hull citadel, broken down into an upper and lower belt that was 12.1 inches deep and sloped at 19 degrees. The sloping of the upper belt was intended to protect the battleship from plunging fire at long range. (*Vladimir Yakubov*)

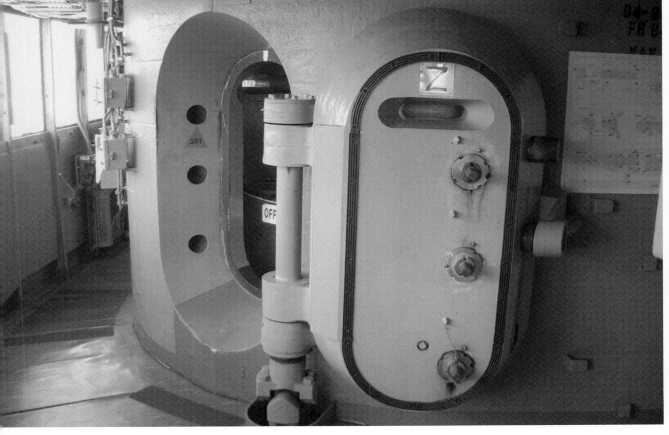

(*Above*) The USS *New Jersey*, as with USS *Iowa* and USS *Wisconsin*, was decommissioned by 1949 and placed into storage. The three decommissioned *Iowa*-class battleships were brought back into service during the Korean War (1950–53), with the primary job of shore bombardment. At the conclusion of that conflict, all four *Iowa*-class battleships were decommissioned and placed in storage. Only the *New Jersey*, seen here in March 1969, was reactivated for use during the Vietnam War (1965–75). The fear of anti-ship missiles resulted in the addition of the large box-like structure below the Mk 38 main battery director, which contained an electronic jammer. (*US Navy*)

(*Opposite*) A view inside the ship control station of the armoured conning tower on a preserved *Iowa*-class battleship. Visible in the centre of the photograph is the engine telegraph. On the left-hand side of the picture is the bottom portion of a periscope jutting down from the compartment's ceiling. The ship control station in the tower contains duplicates of all the equipment found on the navigation bridge of the ship. Located above the ship control station was a fire-control station. (*Vladimir Yakubov*)

The USS *Missouri* is seen here in New York Harbor in October 1945 for fleet week. It was the third ship in the *Iowa* class and was built at the New York Navy Yard, Brooklyn. On 11 April 1945 a kamikaze crashed into the main deck of the ship on the starboard side but only caused a dent in the battleship's armour. The battleship's best-known action during the Second World War was to act as the floating platform upon which the Japanese government signed the official act of surrender on 2 September 1945, bringing the worldwide conflict to its conclusion. (*National Archives*)

Shown in the 1980s is the reactivated USS *Wisconsin* that was constructed at the Philadelphia Navy Yard, Pennsylvania. The battleship joined the fighting in the Pacific theatre of operations during the Second World War, in December 1944. During its combat employment in the Korean War it was struck by a single large enemy artillery round on a starboard 40mm anti-aircraft gun mount, wounding three men. The large antenna array seen on the forecastle of the battleship was for the Naval Tactical Data System (NTDS). (*US Navy*)

Notes

Notes

Notes